ELECTRICAL VEHICLES ENGERGY BALANCE OPTIMIZATION TECHNIQUES ZQ METHOD

PRAKASH GUPTA

TABLE OF CONTENTS

Chapter 4. Active Cell Balancing and Balancing methods Comparative Study for Lithium-ion batteries in Electric Vehicles 67

Abstract

Renewable energy based power systems are the demanded alternative to fossil fuels and greenhouse gas based energy resources over the decades. The evolution of such systems includes varied theories and applications in and over the millennium. Electric Vehicles (EVs) today are one such phenomenal application that addresses the purpose. Lithium-ion (Li-ion) battery-based renewable power systems are proven a great alternative to non-renewable energy-based technologies.

Li-ion batteries prevail in today's portable devices, EVs, and Hybrid EVs. Li-ion batteries are finding more applications because of their high energy and power density, low maintenance, high cell voltage, and less self-discharge. Although they have limitations, they are fragile and need a protection circuit for safe operation. These circuits limit the cell's peak charge voltage and prevent the cell from over discharge. The cell's temperature must be monitored to avoid temperature extremes, and the charge and discharge current should be within a safety limit.

Implementing the Battery Management System (BMS) enables the Li-ion battery to operate within the Safe Operating Area (SOA). Among the various functions of BMS, cell balancing is the important one to focus on. While charging and discharging the battery, all the cells are not charged and discharged equally due to the impedance variations, self-discharge, and temperature characteristics even though the cells are picked from the same manufacturer and the same production batch. The cell balancing mechanism is meant to extend battery life and safety. Cell balancing systems are broadly classified as passive and active cell balancing. In the battery pack, the passive system uses balancing resistors to dissipate the extra charges from the overcharged cells to balance the cell voltages equally. The active cell balancing system transfers extra charge from highly charged to low-charged cells to conserve energy in the battery pack.

An intelligent optimized ML based passive cell balancing scheme is presented in this work. To select a better balancing scheme, an enhanced passive cell balancing is implemented through model-based simulations and circuit implementations. The balancing is carried out, when the EV is in off condition and when the battery is being charged. It is then compared against widely used active inductance based balancing for E-vehicle segments based on the quantum of imbalance, energy efficiency, cost estimation, and balancing speed through simulation. In the simulation, Equivalent Circuit Model (ECM) is implemented, and parameters of the model are acquired using Electro-chemical Impedance Spectroscopy (EIS) test. Final voltage-based algorithm is used to balance the cells which are all at a high State of Charge (SOC). Theoretical investigations are compared with experimental outcomes. Based on the experimental result, the balancing system's important characteristics were implemented by considering energy loss and battery performance.

The major limiting factors of passive balancers are energy loss, temperature rise, and balancing speed. Longer balancing time is acceptable if the EVs are charged through a conventional charging system, and the degree of imbalance is minimum. Fixed resistor passive balancing is not optimal under maximum imbalance conditions and fast charging scenarios. Under fast or slow charging conditions, the variable resistor based passive cell balancing is implemented concerning the balancing current requirement. The parameters considered for the balancing include battery temperature, charging current, cell voltages, and historical balancing records. The balancing resistor selection is made through the ML algorithm. Under different degrees of cell imbalance, temperature, and C-rate, the passive balancing system outcomes are benchmarked, and the system performance is evaluated using ML algorithms, which include Radial Basis Neural Network (RBNN), Back Propagation Neural Network (BPNN), and Long Short Term Memory (LSTM).

Keywords: Electric vehicle, Battery management system, Li-ion battery, Equivalent circuit model, EIS test, Cell balancing, Energy efficiency, Optimization, Machine learning.

CHAPTER 1

Introduction

1.1 Introduction

This chapter details EVs, various types of Li-ion battery chemistry and their cell structure, BMS to monitor and protect the battery, followed by the motivation of this work. Literature surveys on various balancing techniques, balancing optimization, and ML algorithms in BMS are presented. The thesis's objective, organization, and contribution are presented at the end of this section.

1.2 Electric Vehicles

The reduction in the availability of gasoline products and the increase in asset requests are subsequent outcomes for EVs. EVs play a significant role in minimizing the toxins concentration in urban and rural areas. It is a clean alternative to traditional fossil fuel-based vehicles. Conventional internal combustion engine-based vehicles generate power by burning fuels and gases, unlike EVs, which operate on an electric motor. The concept of EVs was introduced long back; however, it has drawn attention in the past decade due to the rising carbon footprint and added environmental impacts [Zakaria, 2019]. Presently, EVs are the right alternative for the environment friendly public and personal transportation. The need for EVs includes the following

- To reduce greenhouse gas emissions

- Finite availability of fossil fuel

- Reduce fuel cost

- Become energy independent

The trend towards more EVs demands high efficiency, high voltage, a long life, and highly safe battery systems.

1.2.1 Li-ion batteries

Batteries are the essential component of EVs, which are used to power the propulsion of vehicles. Enhanced fast charging and improved driving range characteristics are the key objectives to accomplish transport efficiency. The parameters considered for the balancing include battery temperature, charging current, cell voltages, and historical balancing records.

The performance of the EVs is greatly influenced by the batteries, mainly defining the driving range. Batteries are designed with high ampere capacity and energy density to enhance EV's energy and power capacity. Various battery chemistries, such as Lead-acid, Nickel-cadmium, Li-ion and Nickel-metal hydride, etc., which are used in EVs, are compared with specific and volumetric energy density in Fig. 1.1. Among that, the Li-ion variant offers high-energy efficiency, high power-weight ratio, less maintenance, less self-discharge, and no memory effect meeting the requirements of the modern electric vehicle.

Meanwhile, the chemistry of the Li-ion battery is susceptible to overcharge and deep discharge, damaging the battery, life span reduction, and even triggering safety hazards. Consequently, the selection of battery technology and its efficient utilization is of most importance. Beyond energy saving, it is equally important to extend the system's life.

1.2.1.1 Classification

The six of most common Li-ion batteries are as follows

- Lithium Manganese Oxide (LMO)

- Lithium Cobalt Oxide (LCO)

- Lithium Nickel Manganese Cobalt Oxide (NMC)

- Lithium Iron Phosphate (LFP)

- Lithium Nickel Cobalt Aluminium Oxide (NCA)

- Lithium Titanate (LTO)

Li-ion battery variants are compared concerning specific power, energy, performance, cost, life span, and safety which is represented in Fig. 1.2. The LCO, NMC, and NCA chemistry provide high specific energy, whereas LFP and LTO are the safest and long cycle life Li-ion battery chemistry compared to other variants. In terms of thermal stability, LMO is superior. The cycle life and safety will dominate over capacity, especially in electric power trains. Hence, the right choice of battery chemistry is based on the user requirement and the type of application.

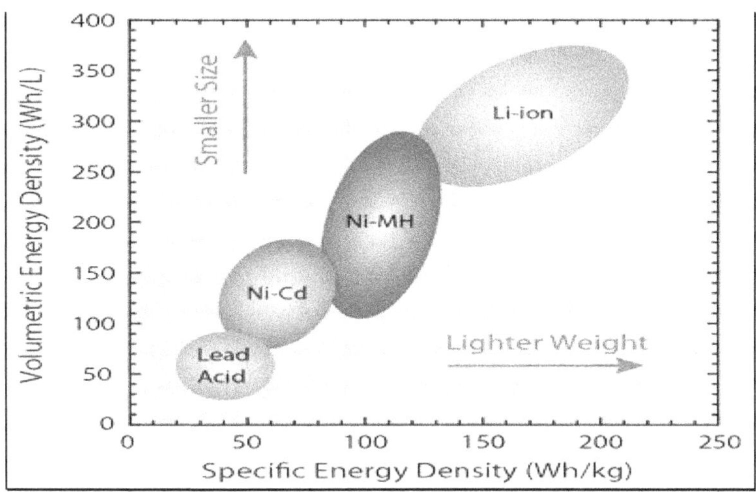

Figure 1.1 EV battery comparison [Epectec]

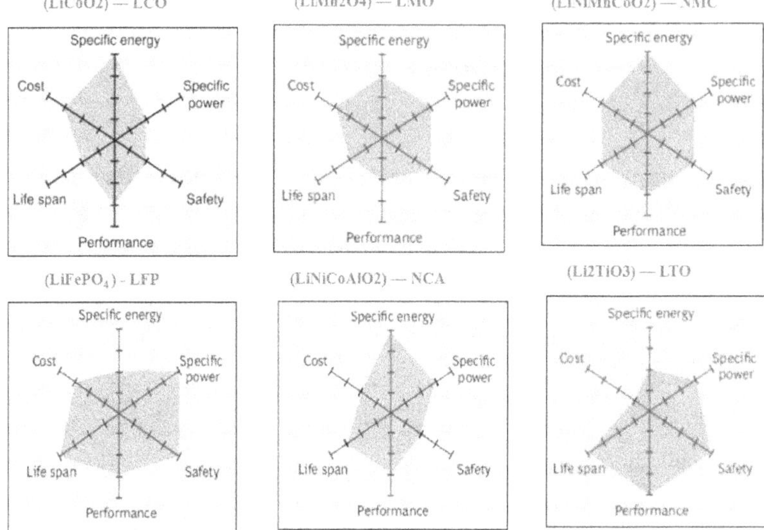

Figure 1.2 Li-ion chemistry comparison [batteryuniversity]

Increasing demand vs the availability of Li-ion ore seeks increased cell life span through appropriate battery management systems. To design the right BMS, Li-ion battery characteristics have to be studied and modelled thoroughly.

1.2.1.2 Li-ion cell structure

Each Li-ion cell has the following essential components cathode, anode, separator, and electrolyte. While charging or discharging the batteries, reduction-oxidation occurs in the anode and cathode electrodes. The porous membrane acts as a separator that isolates the positive and negative electrodes electrically and allows Li+ ions to moves across the electrodes. Li salt (LiPF6) which is dissolved in an organic solvent acting as an electrolyte. Fig.1.3 shows the Li-ion cell composition.

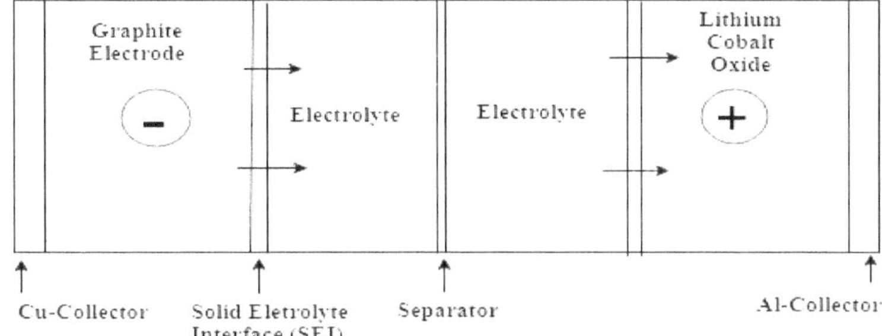

Figure 1.3 Li- ion cell composition [Saidani, 2017]

During discharge, oxidation takes place in the anode.

$$x.Li.C_6 \rightarrow x.e^- + x.Li^+ + x.C_6 \tag{1.1}$$

Reduction takes place in the cathode.

$$x.e^- + Li_xCoO_2 + x.Li^+ \rightarrow LiCoO_2 \tag{1.2}$$

For example, while charging in an LCO, the Li-ion (Li+) passes through the electrolyte from the cathode to the anode. Through the outer circuit, in the same means electrons (e^-) travel from the cathode to the anode. The Li^+ accumulates between the layers of graphite. The process, as mentioned above, is reversed under a discharge regime.

1.2.2 Battery management system

Li-ion battery needs an implementation of a suitable BMS to keep every cell within its reliable and safe operating region. The key objectives common to all BMS consists of

- Save the battery from abuse and damage
- Extending the battery life cycle
- Ensuring that the battery is always available for use

Factors monitored and controlled by the BMS include [Andrea, 2010].

- Supply voltage
- Cell or battery voltage
- Discharging and charging rates
- Temperature measurement and control
- State of charge/health calculation

The overall BMS includes the following parts: management, monitoring, balancing and protection. Among these, balancing is essential in battery life since cell voltages will drift over time without balancing. The total capacity of the battery pack will reduce rapidly during operation, hence the failure of the entire battery framework and a reduction in battery life. This condition is severe for a high voltage battery system and is frequently charged through regenerative braking (Hybrid EVs). Therefore, the BMS requires an optimal balancing mechanism to yield the following:

- Higher state of charge

- Improvising the battery pack's performance

- Reduction in cell degration

- Thermal runaway avoidance

- Extension of drive cycle and life of the cell

- Battery safety enhancement

BMS adapts intelligent monitoring and tuning mechanics, possibly using the microcontrollers (to run the finalized algorithm), sensors, A/D converter with a suitable feedback circuit to maintain the battery system within the safe operating area.

1.3 Motivation for the work

In the EV battery pack, the individual cell voltage diverges over time due to the difference in the SOC, impedance, capacity, self-discharge rate, and temperature characteristics. Cell balancing is not only increasing the lifetime and performance of the battery but also enhances battery safety. The passive balancing technique is simple and easy to implement; excess energy from the overcharged cell is dissipated through the bypass route. It offers a relatively low cost solution for balancing the cells widely used in industrial applications even though energy is dissipated as heat. Active balancing uses capacitive or inductive shuttling to move the charge from overcharged to the undercharged cell with significant efficiency instead of being bled off. However, there is a trade-off between improved efficiency and more components with additional system complexity.

The key objective of this work is to improve the minimize energy loss, faster balancing, maximize the capacity and temperature suppression rate of the passive balancing system under fast charging scenario. This can be achieved through tuning and optimization of the passive algorithm for efficient thermal control resistor in shunt. It distributes balancing act across a window with appropriate idle time in between, which can be catalogued for various use case scenarios using the ML approach.

1.4 Literature Survey

Various cell-balancing approaches are studied and analysed effectively. The reduction of power loss and balancing time in the passive system is discussed in many literatures through control algorithms. The ML algorithms has revolutionized through the usage of data-driven approach and the electric vehicle BMS domain to improve the system measurements' accuracy and learning capabilities. The following sub-sections present the summary of these approaches

1.4.1 Cell balancing approaches

The significance of cell balancing is to balance the SOC and voltage between the cells when they are fully charged or discharged [Omariba, 2019]. The balancing algorithms detect and control the cell imbalances, classified as SOC, OCV, and terminal voltage based [Ma, 2018]. Using SOC or OCV provides better SOC estimation; however, the result depends on the SOC estimation algorithm's accuracy [Lin, 2020]. It is not an efficient approach for batteries with SOC curves versus flat OCV (e.g., LiFePO4). The balancing method using data-driven approach SOC estimation showing significant results but needs in-depth domain knowledge and large quanta of data [How, 2019]. The least complexity implementation approach based on voltage are extensively used in E-mobility systems [Tavakoli, 2020]. In BEV, at the final stage of the charging process in balancing is frequently done using voltage-based methods [Plett, 2004]. Final voltage algorithm balances in less time hence needs higher balancing current level and it works at the end of charging [Chen, 2018]. It avoids working in the flat portion of the SOC versus OCV curve as represented in Fig.1.4.

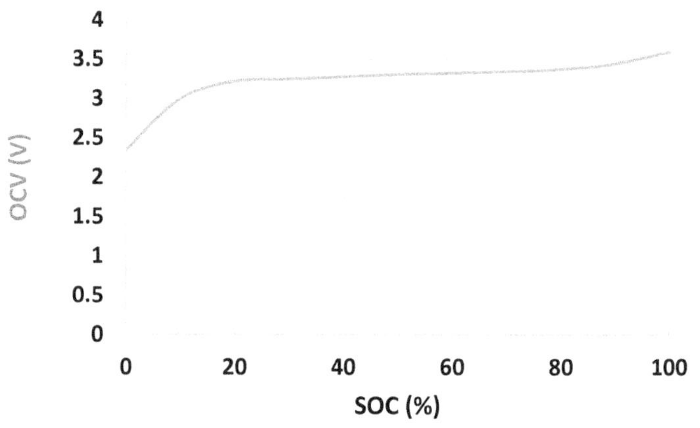

Figure 1.4 Lithium ion phosphate battery SOC Vs OCV characteristics [Ma, 2016]

Errors occurring due to internal resistance is minimized by reducing the charging current. The intelligent algorithms (e.g., neural network, fuzzy, graph theory, and genetic algorithm) are used in the balancing to achieve stability, accuracy and transition time [Wang, 2020], [Zhang, 2015], [Shan, 2020]. However, the system complexity increases when more cells are added.

The module level balancing [Han, 2019] is hierarchical in with respect to pack, module, and cell level, which is effective for large-scale energy segments. [Han, 2019] proposed a dynamic reconfiguration-based emerging balancing technique, capable of matching the user behaviour and requirements by changing the battery interconnection. An optimal control approach [Ouyang, 2020], [Docimo, 2019] accomplishes the key cell balancing objectives such as reduction in energy dissipation, shortening balancing time, increasing efficiency and capacity, and reducing the temperature raise. BMS with passive balancing for Li-ion batteries replacing lead-acid batteries in an electric kart is designed by [Vitols, 2014] to increase the pack's current capability and energy/mass ratio. [Amin, 2017] developed a passive balancer for 15 series connected cells by using internal MOSFET resistor for balancing, reduction in BMS hardware size, and improvement in balancing current to 3.07A, hence reduction in balancing time is seen. However, for the higher imbalance scenarios, it takes more time to balance. The dissipative shunt resistor is replaced with MOSFET [Xu, 2019] to enhance the performance while reducing the system cost. The amplifier mode of MOSFET is triggered during the imbalance and the excess energy of the highly charged cell is dissipated through it.

It is possible to regulate the balancing current up to 1.2 A. The frequency of the balancing switch ON/OFF from 1,150 to 2 cycles and the SOC variance to 0.157 are both greatly reduced by the outlier detection-based balancing method mentioned in [Piao, 2015]. The algorithm can accurately forecast the abnormal cell, however some alterations in the voltage cut-off value have been noted. Despite the high circuit cost and complexity, [Schmid, 2017] investigated a passive balancing architecture that balances the cells electrochemically without the addition of any electrical devices. According to Valchev (2018), balancing at various charging currents (such as 1C, 0.5C, etc.) reduces the balancing time by 20%. The main components for dissipating the extra energy are power transistors and power resistors [Perisoara, 2018]. The power dissipation across each transistor and resistor, however, is roughly 7.25W and 3.6W, respectively. It makes the balancing process less effective. An algorithm for balancing was suggested by the author [Pichon, 2015] and is based on SOC and cell capacity. However, evaluating the capacity and SOC precisely is a challenging process that calls for a sophisticated BMS. For high power battery packs, the cell balancing method described in [Probstl, 2018] reduces cell ageing by up to 23.5% using passive balancing and 17.6% using active balancing while taking into account the lowest load current to the cells with a low State of Health (SOH). As a result, the low SOH cell life is increased, lengthening the lifespan of the entire battery pack. The justification for choosing the passive balancing system is highlighted by [Samaddar, 2021] discussing the trade-offs (expensive to implement, complex control, etc.) of using an active balancing system.

The single switched capacitor based active cell balancing system presented in [Daowd, 2012] reduces the system cost but takes more time to balance the cells, which have minimum ΔV. [Daowd, 2013] suggests using switched capacitor balancing to transmit energy between nearby cells. In order to drastically reduce the balancing time, a modularized balancing system using a single switched capacitor for cell-level balancing is modelled and prototyped in [Daowd, 2014]. To reduce balancing time, chain structured switched capacitors [Kim, 2014] and series-parallel switched capacitors [Ye, 2015] are introduced. The Double-Tiered Switched Capacitor (DTSC) is created by adding an additional capacitor tier to the switched capacitor topology, and the slower balancing speed of the DTSC is overcome by employing the resonant topology in [Ye, 2017]. [Shang, 2019] discusses optimised modularized switched-capacitor-based methodologies, such as shuttling capacitor methods.

Inductor based scheme offers good performance with a trade-off between implementation cost and balancing speed [Xie, 2011]. A next-to-next inductor balancing topology introduced by [Phung, 2014] transfers energy among adjacent cells, consisting of (n+2) switches, 'n' inductors, and capacitors for 'n' cells. It prolongs the driving range, but the higher number of components increases the system's complexity. The long balancing time of inductor-based balancing topology is reduced by fixing independent balancers in different layers [Dong, 2015]. The author [Zhou, 2016] introduced any cell to any cell topology to enhance the balancing current and efficiency, but more switching elements increase the system size and cost. In a single inductor based topology, energy is transferred from over-charged cell to under-charged cell, similar to buck-boost techniques [Vardhan, 2017]. The coupled inductor topology introduced in [Moghaddam, 2019] shows excellent balancing characteristics with fewer components reducing the cost further than the traditional one [Xie, 2011]. Two-stage multi inductor based topology introduced by [Ding, 2020] in which energy transfer takes place from high SOC to low SOC cells directly. The balancing time is reduced effectively, and it took 10 min for 3% ΔSOC.

The single transformer cell topologies balance the charge among cells using only one transformer with high balancing speed and low magnetic losses [Moghaddam, 2018], [Shang, 2017]. Balancing topology based on a single transformer per cell proposed by [Conway, 2021] can monitor and balance the cell with a current of 50 mA to 0.5A. Hence, there is no need for separate isolation because the transformer itself acts like an optocoupler. The SOC mismatch is the only factor that influences the balancing time; it is not depending on the number of cells in the pack.

Converter based topologies offer better efficiency and good control over power flow [Amjadi, 2010]. The buck-boost topology proposed by [Shubiao, 2017] is applicable where the imbalances are the minimum, and rapid balancing is required. Zero voltage and zero current switching quasi-resonant converter is an efficient converter in which the capacitor and inductors are added together to make a resonant tank to achieve zero current switching [Shang, 2015]. In forward and fly-back converters, the excess energy from the overcharged cell is transferred to low-charge cells through transformer winding, switches, and diodes [Shang, 2017]. Cuk converter-based topology using coupled inductor is proposed by [Moghaddam,

2019] to balance only the neighboring cells. DC-DC converter based system for increasing the life cycle of series connected battery packs [Turksoy, 2020] shows high balancing speed (< 20 min) with respect to other topologies.

The additional parts required to implement the active system raise the total cost of the BMS. In contrast, these active circuits' standby power losses are greater than those of the passive system [Andrea, 2010]. Active balancing systems are effective and rapid at balancing, but their cost, circuit complexity, and dependability raise questions about whether their advantages are justified. The passive balancing system is widely employed in industrial applications as a result of these factors [Ditsworth, 2018].

1.4.2 Machine learning approaches

According to [Liu, 2021], [Ardeshiri, 2020], and [Man-Fai, 2020], machine learning techniques are among the most effective methods for estimating critical battery performance indicators like the Remaining Useful Life (RUL), SOC, and SOH of the batteries. These techniques also increase the availability of battery data and increase power-computing capacity. [Vidal, 2020] compares various machine learning (ML) techniques, including feed-forward neural networks (FNNs), radial basis functions neural networks (RBNNs), support vector machines (SVMs), and recurrent neural networks (RNNs), in terms of battery types, system accuracy, inputs and outputs, data quality, and test scenarios. The robust algorithm of the neural network can manage any complicated non-linear system under various dynamic loads and temperatures [Chaoui, 2017]. In [Cui, 2020], an evolutionary control method is suggested for the best battery pack design to address issues in the EV arena. By utilising the backtracking search approach for estimating battery status, the authors [Hannan, 2018] have presented an offline optimization strategy for the best FNN structure. A reinforcement learning architecture is put forward in [Sui, 2020], and learning agents are instructed to train themselves to balance the unbalanced cells. To assess the battery parameter for an accurate estimation of SOC, a high current pulse injection approach is integrated into the balancing circuit using a FNN-based algorithm [Wang, 2019]. RBNNs are frequently described as having very quick online learning speed, high input noise tolerance, and effective interpolation [Yang, 2019]. An average pack level model is sufficient for better voltage prediction in Li-ion batteries because the parameter uncertainties.

To predict the battery SOC and voltage of an electric motorcycle, [Caliwag, 2019] presents an LSTM and a hybrid vector autoregressive moving average (VARMA). The authors of [Vidal, 2019] have put out an original idea for using an LSTM with transfer learning to shorten training time. Different optimization methods, including Gauss-Newton, Levenberg-Marquardt, and steepest descent, are applied to Li-ion battery technology in [Ramadesigan, 2012] to address problems like under-utilization, capacity fading, and material damage as a result of stress from a system engineering perspective. The approach described in [Gopalakrishnan, 2019] makes use of an ideal layer design while taking thermal constraints into consideration to lessen the possibility of lithium plating during fast charging scenarios. It also explains how, at greater charging temperatures, the thermal system design limits vehicle speed. However, the previous work did not emphasize the ML approach for cell balancing.

Key research gaps identified in the balancing systems are as follows

> - Heat loss from the balancing resistor in the passive balancing system causes thermal problems and reduces battery performance.
> - Passive cell balancing takes more time to balance. It is efficient during charging mode.
> - Since a passive system takes more time to balance, under a high degree of imbalance DC fast charging and extremely fast charging technology is not a feasible option.

> - The high cost and circuit complexity of the active balancing system are the main barriers for the real time usage

1.5 Objectives

> ➢ Select an appropriate battery modelling method to predict the charge/discharge characteristics, state of charge, voltage, and other battery parameters.
> ➢ To conduct an analysis of the cost, energy efficiency, and power loss for passive and active balancing systems, which enhance battery performance.
> ➢ To use an ideal balancing strategy to increase the speed of the passive balancing system.
> ➢ To introduce machine learning based efficient and novel control algorithms to enhance battery life while optimally reducing the balancing time and system temperature.

1.6 Contribution of the thesis

Below are the key aspects considered as a part of this thesis:

> ➢ Proposal of a better battery model for the EV application by comparing the battery model outcomes with experimental data. The feasibility analysis is done for the selected ECM model, and the model parameters are obtained by conducting an EIS test on the Li-ion cell.
> ➢ Implementing passive and active cell balancing systems for E-mobility application by considering the most significant characteristics (e.g., energy efficiency, thermal behaviour, and balancing time) of the balancing system through simulation.
> ➢ A passive balancing system is implemented in hardware to confirm the results of the theoretical inquiry into power loss and energy efficiency.
> ➢ Based on the degree of imbalance, balancing speed, battery efficiency, temperature, and system cost, the passive balancing architecture is contrasted with the widely utilised inductor-based balancing topology for various e-vehicle sectors.
> ➢ The use of a passive cell balancing system based on machine learning (ML) with a variable balancing resistor under slow and fast charging conditions

to increase balancing speed and address thermal problems to prolong battery life.

1.7 Organization of the thesis

Chapter 2, details the state of art techniques in battery modelling methods that are applicable in the EVs sector, EIS test to find out the model parameters, battery model comparison, and comparison of simulation outcomes of RC models with respect to experimental data are discussed.

Chapter 3, mainly focuses on the simulation and hardware implementation of passive cell balancing for 48V Li-ion battery. Battery pack design, balancing implementation in the BMS system, hardware realization, and passive system outcomes are explored.

Chapter 4, analyses the various active cell balancing scheme with respect to cost, size, speed, circuit design, and efficiency. Based on the analysis an inductor based active balancing topology is implemented in MATLAB. Simulation outcomes are discussed and compared with respect to a passive balancing system to select a suitable balancing scheme for various EV segments.

Chapter 5, the optimization strategy for enhancing the passive system's balancing speed in both traditional and fast charging conditions is covered. The passive balancing system with changeable resistor outcome is investigated using MATLAB/Simscape. The balancing time, speed, and power loss of the cell at various delta voltages and C-rates during the charging time are the parameters considered for the analysis. Based on the differences in voltage between the cells, balancing performance is examined. The suggested balancing system uses an ML algorithm for optimising resistors. The suggested ML-based passive balancing system is compared in this chapter to the results of the traditional passive balancing system.

Chapter 6, presents the conclusion of the work with future research orientations.

1.8 Conclusion

This chapter discusses the importance and needs of EVs for the present-day scenario, a comparison of EV batteries, cell structure, and the importance of the BMS system. The various internal and external factors cause the cell voltage imbalance within the pack, which degrades the performance and leads to thermal runaway. To increase battery life and safety, the performance of the balancing system must be improved with an appropriate balancing control algorithm. A brief literature survey is given for different cell balancing techniques to improve the balancing performance and discusses related work in ML and passive balancing. To improve the balancing time and speed, a literature review on cell balancing optimization is discussed. Additionally, the thesis's purpose, goals, and contribution are discussed.

CHAPTER 2

Battery Modelling and Parameter Estimation from Electrochemical Impedance Spectroscopy test

2.1 Introduction

This chapter analyses the widely used equivalent circuit models through simulation. The findings are compared with the experimental data to select a suitable model to predict the charge/discharge characteristics, SOC, and other battery parameters. EIS test is conducted to determine the Li-ion cell impedance characteristics at wide frequency spectra. With regard to SOC, a parameter estimation algorithm that uses EIS data as an input customizes the model parameters.

2.2 Battery modeling

To develop an effective battery-operated system, a precise battery model is necessary. It has the ability to forecast the battery's dynamic and steady state behaviour. The input and output parameters of the battery model is shown in Fig. 2.1.

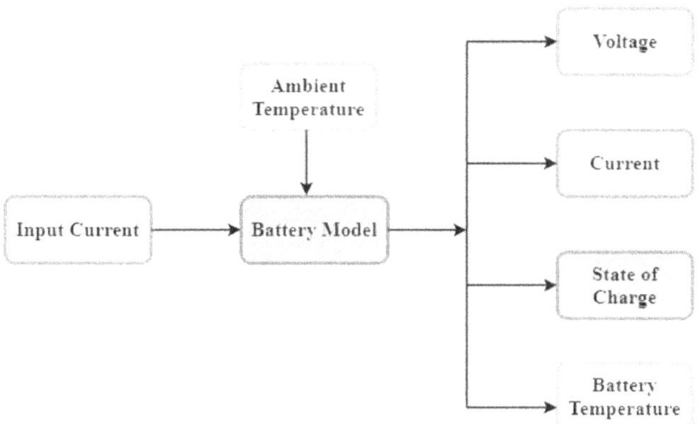

Figure 2.1 Battery model input and output parameters [Tremblay, 2009]

17

The categorization of the battery modelling is shown in Fig. 2.2 based on the SOC measurement [Meng, 2018] and [Fotouhi, 2016]. Using a set of battery data and a variety of elements, data-driven and model-based approaches try to properly forecast the battery performance characteristics with the use of mathematical equations and intricate algorithms [How, 2019].

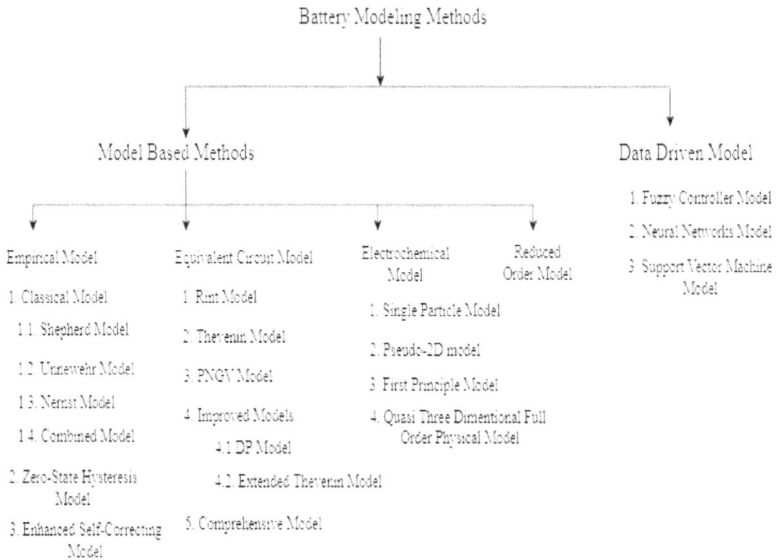

Figure 2.2 Battery modelling methods

2.2.1 Model-based methods

The SOC, voltage, and other battery parameters can be determined effectively and precisely using model-based methods. It is made up of an algorithm to estimate the battery parameter and a battery model. The model's inputs, including the outside temperature and current, are further examined, and the model's outputs are measured and compared to the most recent battery data. By fine-tuning the model parameters to increase performance, error in the model output is reduced.

2.2.1.1 Empirical model

Black box or mathematical models are other names for empirical models. Transfer functions are used to derive the output from the input parameter using a straightforward mathematical method. Although it has a limited degree of precision, it responds quickly and is simple to configure the model parameters. These models fall into three categories: zero-state hysteresis models, enhanced self-correcting models, and classical models.

The electrochemical models, such as the Nernst, Shepherd, Unnewehr, and combined models, are simplified to create the classical model (combination of all). The model's output voltage response depends on the battery's state of charge (SOC) and current. Other higher-order complex battery models do a poorer job of describing the non-linear response of the battery than a straightforward set of mathematical and algebraic equations. To improve the accuracy of the combined empirical model, a modified empirical model was developed that uses the shepherd equation to take into account the battery charge and discharge cycles separately.

The hysteresis effect is taken into account in the zero-state hysteresis model, but it is not in the classical model [Plett, 2015]. After the battery has been discharged and recharged, the hysteresis has an impact on the SOC calculation. For more accurate SOC estimation, the hysteresis effect should be taken into account. The diffusion voltage decays to zero as soon as the battery is allowed to rest since it is time-dependent, but the hysteresis voltage is dependent on the battery SOC. Because it accommodates ohmic losses, hysteresis, and polarisation time constants, an enhanced self-correcting model, which goes by the name enhanced, may accurately capture battery dynamics. It is self-correcting because, when the cell is at rest, the model voltage converges to OCV and hysteresis, and when the current is constant, it converges to OCV plus hysteresis less the voltage drop across the resistance components [Plett, 2015]. It may be a superior model for electric vehicle BMS because it takes into account the relaxation impact in the EV application.

2.2.1.2 Equivalent circuit model

By using simple battery models, the BMS has to predict the steady and dynamic characteristics of the battery accurately in real time. However, simultaneously, it is challenging to simulate the dynamic behaviour of the battery by using a simple circuit model. However, complex models are more challenging to implement for real-time applications. It does this by substituting the comparable straightforward electrical circuit, as depicted in Fig. 2.3 [Westerhoff, 2016], with the complex electrochemical process found inside the battery. As a result of its ease of use for real-time applications, it is a popular model for BMS for electric vehicles.

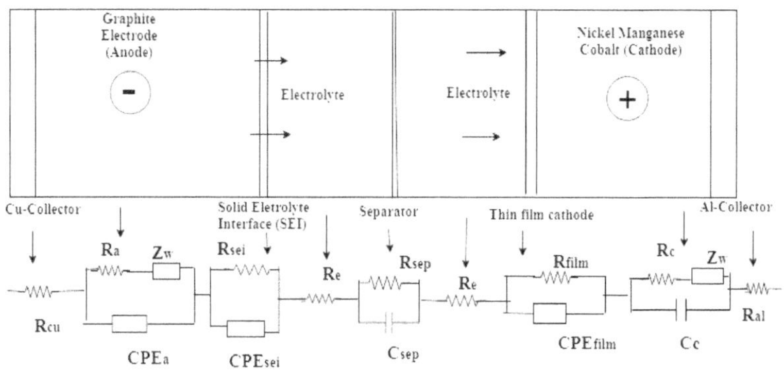

Figure 2.3 Equivalent circuit elements represent cell components [Westerhoff, 2016].

The internal resistance of the electrodes, electrolyte, current collectors, and separator, among other components, is indicated by the resistor in the ECM. The double layer, charge transfer impact of the electrolyte and electrodes are reflected in the diffusion behaviour and Warburg impedance (Zw) and Constant Phase Element (CPE) of the parallel RC network. In the literature on ECMs, the Rint, Thevenin, Partnership for a New Generation of Vehicles (PNGV), and expanded Thevenin models are well known. In Fig.2.4, various ECMs are depicted.

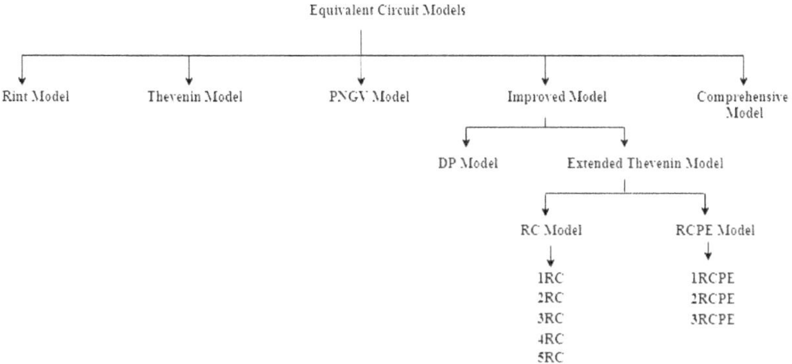

Figure 2.4 Equivalent circuit models

The ECM is an accurate model to predict the battery operating characteristics with battery SOC, voltage, current, and temperature with the help of a simple set of equivalent electrical circuit elements [Saldana, 2019]. Fig 2.5 represents the circuit diagram of the model-based methods.

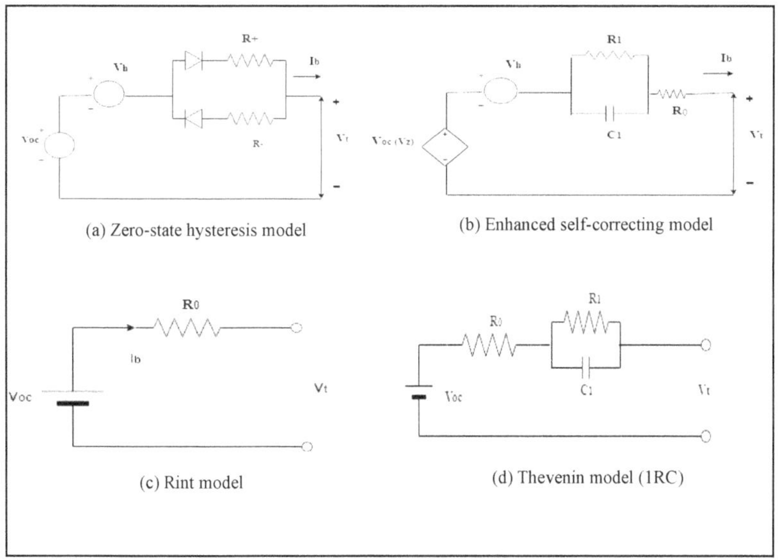

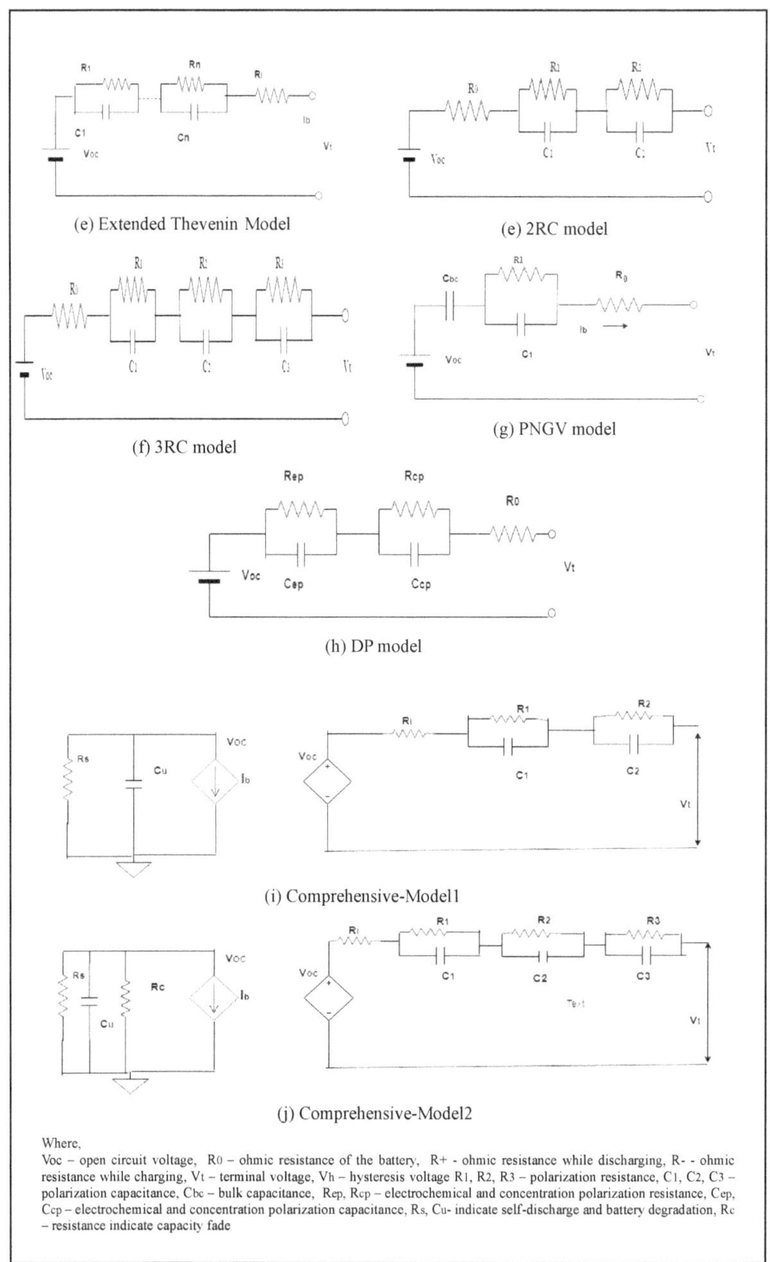

Where,
Voc – open circuit voltage, R0 – ohmic resistance of the battery, R+ - ohmic resistance while discharging, R- - ohmic resistance while charging, Vt – terminal voltage, Vh – hysteresis voltage R1, R2, R3 – polarization resistance, C1, C2, C3 – polarization capacitance, Cbc – bulk capacitance, Rep, Rcp – electrochemical and concentration polarization resistance, Cep, Ccp – electrochemical and concentration polarization capacitance, Rs, Cu- indicate self-discharge and battery degradation, Rc – resistance indicate capacity fade

Figure 2.5 Circuit diagrams of model-based methods

The Rint model is a straightforward R model that is used to describe the battery's steady-state behaviour, but it is unable to account for the battery's transient reaction to changing load conditions [He, 2011]. The Thevenin model, a parallel RC pair that captures the nonlinear response of the battery, is incorporated into the ECM. By significantly enhancing the Thevenin model circuit components, including an additional capacitor, the PNGV model is created. The capacity and DC responsiveness of the battery are described [Li, 2014].

An improved Double Polarization (DP) model is obtained by including another parallel RC pair in the Thevenin model, representing the dual-polarization effect [He, 2011]. The polarization effect refers to the displacement of electrode potential from the equilibrium value, which reduces the performance of the battery. DP model includes the electrochemical and concentration polarization effect of the battery.

In the extended Thevenin model, to improve the model's accuracy, more parallel RC and RCPE pairs are added. Practically, it is challenging to design an exact model for various battery chemistries due to the complex electrochemical behaviour of the batteries. Based on the system's accuracy and implementation complexity, the extended Thevenin model can capture the battery transients in various levels such as 1RC, 2RC, and 3RC up to the nth order model. Hence, the RC models are implemented to benefit the EV's BMS. The battery models and their characteristic equations [Li, 2014], [Plett, 2004] are described in Table 2.1.

Table 2.1 Battery model equations

S. N	Battery Models	Equations
		Empirical Model
1	Shepherd model [Hussein, 2011]	$V(t) = V_0 - R_0\, i(t) - \dfrac{K_1}{Q(t)}$ *Where,* $V(t), V_0$ - *output voltage & OCV of the battery* $i(t)$ – *Battery current.* R_0 – *Ohmic resistance* K_1 - *constant used for curve fitting,*

		$Q(t)$ – battery SOC
2	Unnewehr model [Meng, 2018]	$V(t) = V_0 - R_0\, i(t) - K_2 Q(t)$
3	Nernst model [Plett, 2004]	$V(t) = V_0 - R_0\, i(t) + K_3 ln(Q(t)) +$ $K_4\ ln(1\text{-}Q(t))$
4.	Combined model [Tremblay, 2009]	$V_k = K_0 - R_0\, i_k - \dfrac{K_1}{Q_k} - K_2\, Q_k +$ $K_3\ ln(Q_k) + K_4\ ln(1 - Q_k)$ *Where* *k – time index, V_k – battery terminal voltage*
5	Zero-state hysteresis model [Plett, 2015]	$V_t = V_{oc}\,(Z_k) - I_k\, R_0 - S_k\, M - M h_k$ *Where,* *Z_k - describes SOC* *M - constant coefficient describing the hysteresis level* *S_k - represents the sign of the current* *h_k - hysteresis voltage*
6	Enhanced self-correcting model [Plett, 2004]	$V_t = V_{oc}(Z_k) - I_k\, R_0 + S_k\, M + M h_k$ $- \sum R_j\ i_{Rj}$ *Where,* *$R_j i_{Rj}$ – voltage drop across the RC pairs*
		Equivalent circuit model
7	Rint Model [He, 2011]	$V_t = V_{oc} - I_b\, R_0$ *Where,* *I_b – battery current*
8	Thevenin Model [Plett, 2015]	$\dfrac{dV_1}{dt} = \left(\dfrac{I_b}{C_1}\right) - \left(\dfrac{V_1}{R_1\, C_1}\right)$ $V_t = V_{oc} - I_b\, R_0 - V_1$ *Where,* *V_1 – drop across the 1RC network* *$R_1\, C_1$ – time constant of the 1RC network*
9	PNGV Model [Li, 2014]	$\dfrac{dV_1}{dt} = \left(\dfrac{I_b}{C_1}\right) - \left(\dfrac{V_1}{R_1\, C_1}\right)$ $V_t = V_{oc} - I_b\, R_0 - V_1 - V_{bc}$ *Where,* *V_{bc} - drop across the bulk capacitance*
10	DP Model [He, 2011]	$\dfrac{dV_{ep}}{dt} = \left(\dfrac{I_b}{C_{ep}}\right) - \left(\dfrac{V_{ep}}{R_{ep}\, C_{ep}}\right)$ $\dfrac{dV_{cp}}{dt} = \left(\dfrac{I_b}{C_{cp}}\right) - \left(\dfrac{V_{cp}}{R_{cp}\, C_{cp}}\right)$ *Where,* *$R_{ep}\, C_{ep}$ – electrochemical polarization time constant* *$R_{cp}\, C_{cp}$ – concentration polarization time constant*

| 11 | Extended Thevenin Model [Chen, 2006] | $$\frac{dV_1}{dt} = \left(\frac{I_b}{C_1}\right) - \left(\frac{V_1}{R_1\,C_1}\right)$$ $$\frac{dV_n}{dt} = \left(\frac{I_b}{C_n}\right) - \left(\frac{V_n}{R_n\,C_n}\right)$$ $$V_t = V_{oc} - V_1 \dots -V_n - I_b\,R_0$$ *Where,*
 $R_n\,C_n$ – time constant of the nth RC network
 $V_1 \dots -V_n$ – drop across 1st RC to an nth RC network |

More RC networks are added to the Thevenin model to enhance the model accuracy. The number of RC pairs varies based on the type of application, and there is a trade-off between the circuit complexity and accuracy when the levels are increased.

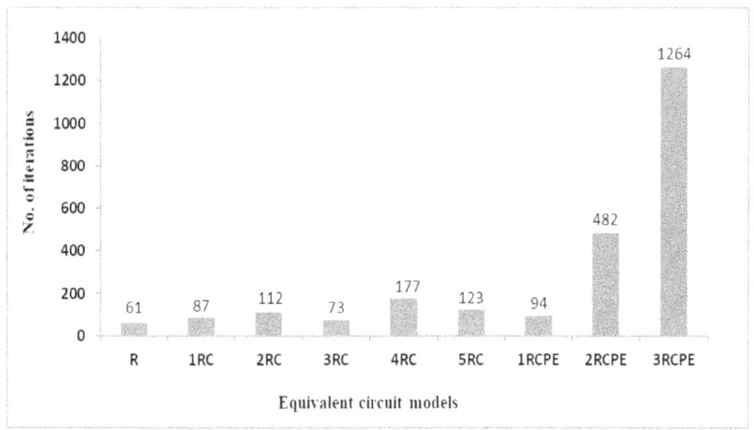

Figure 2.6 Total number of iterations to get optimal parameter [Westerhoff, 2016]

Comparing 3RC to other low and higher order models, Fig. 2.6 above demonstrates how few iterations are needed to obtain the ideal model parameters. The different uses of ECMs are listed in Table 2.2 [Westerhoff, 2016]. The static properties of the cell can be described by a single R model, which is appropriate for grid integration. The 1RC model outlines the features of charge transfer, but the 2RC model also takes the diffusion process into account. Both can be integrated into the smart grid. E-mobility employs 1RCPE and higher-order RC models for more precision. When accuracy is crucial, higher-order RC models with CPE are employed in diagnostic tools.

Table 2.2 Application of ECM

Sl.No	Model name	Application
1	Simple model	High voltage grid integration
2	1RC,2RC	Smart grid integration
3	3RC,4RC,5RC,1RCPE	E-mobility
4	2RCPE,3RCPE	Diagnosis

The run-time model is another name for the comprehensive model. The comprehensive model takes into account self-discharge and battery capacity fading. The battery transient reaction is included in the second portion of the model, which first simulates battery degradation over calendar ageing and cycling [Chen, 2006]. A resistance that replicates capacity fade is introduced to the comprehensive model 2 in order to determine battery ageing and RUL [Li, 2014]. Table 2.3 lists the various ECMs and their primary influencing factors.

Table 2.3 ECM models and key influence factor

Model	Polarization Characteristics	Dual Polarization	Self-discharge	Capacity-fade
Rint	√	X	X	X
Thevenin	√	X	X	X
PNGV	√	X	X	X
DP	√	√	X	X
Extended Thevenin	√	√	X	X
Comprehensive-1	√	√	√	X
Comprehensive-2	√	√	√	√

All ECMs have polarisation properties built in. The comprehensive model 2 effectively incorporates all of a battery's dynamic properties over the course of its lifespan.

2.2.1.3 Electrochemical model

Electrochemical models are called a physical or white-box models. This model describes the complex electrochemical process happening inside the cell with high accuracy [Han, 2015]. It includes more partial differential equations that need to be resolved concurrently. Although it is time consuming, complex, and challenging to use in a real-time application, it is an appropriate model for battery manufacturers.

2.2.1.4 Reduced order model

Electrochemical models need to be simplified to make them compatible with the estimation algorithms implanted on the controller board. The simplified model reduces the computational complexity with a better model performance by considering additional assumptions. Simulation of this model is faster than experimenting in a real-time system, which makes this model a preferred choice for battery state estimation, fault diagnosis, and aging analysis [Fan, 2018].

2.2.2 Data-driven model

Data-driven models are implemented using fuzzy controllers [Singh, 2006], neural networks [Yang, 2019], and support vector machine [Junping, 2006] algorithms and need a lot of data to process. This model uses the output of the learning module, which uses the battery current, voltage, temperature, and SOC as inputs. The output from the learning module is used to estimate the system behavior, which does not require a precise system model. The control approach can provide a more significant benefit where the system is highly non-linear and cannot be modelled using simple mathematical equations.

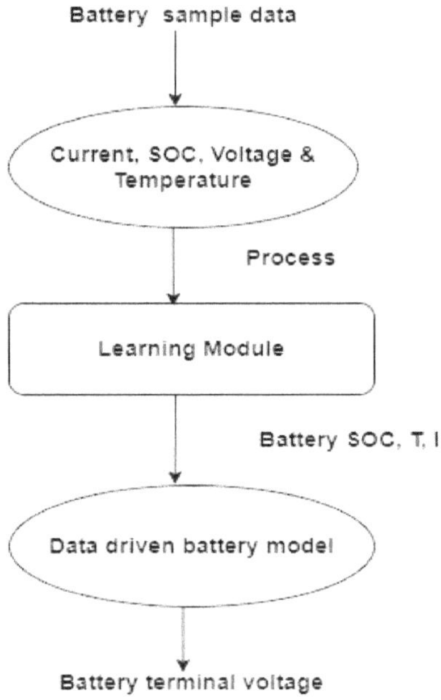

Figure 2.7 Data-driven model [Meng, 2018]

It can be combined with model-based methods to achieve a superior data model. The relationship between the data-driven model's output and input voltage is represented in Fig. 2.7. Table 2.4 compares battery modelling methods.

Table 2.4 Comparative study on battery modelling

Battery Models	Mathematical Model	ECM	Physical Model	Reduced order Model	Data-driven Model
Accuracy	Low	Medium	Very high	High	High
Voltage measurement	Limited capability in the terminal voltage estimation	Parameter estimation is complex for voltage measurement	Very complex and time consuming	Easy to do voltage measurement	The tedious data collection process
Physical interpretability	Low	Limited	High	Medium	Medium

Time	Simple and low time consuming	Medium time consuming due to simple and easy to understand	Takes more time due to accurate voltage calculation	Average time consumption	It takes less time as prior battery knowledge is not needed
Computational Complexity	Low	Medium to low	High	High	High
Configuration Effort	Low	Medium	High	Medium	High
Application	All energy storage system	Real-time monitoring and BMS	Battery system design stage and diagnosis	SOC estimation, control application	Electric vehicle, Hybrid electric vehicle

Each method has its advantages, and the battery model performance is based on environmental conditions, battery-operating temperature, SOC operating range, and battery ageing. The electrochemical model is the base model for all. A simplified electrochemical model is called a mathematical model. ECM is obtained by replacing the chemical process inside the cell with equivalent electrical circuit elements. Data-driven model states the electrochemical characteristics with high accuracy data analysis. The model should be as simple as possible for practical application, but there is a trade-off between model simplicity and accuracy, as represented in Table 2.5.

Table 2.5 Battery model comparison from literature

Models	Merits	Demerits	Literatures
Mathematical Model	Simple and less time consumption	Low accuracy	[Tremblay, 2009], [Hussein, 2011], [Plett,2015], [Meng, 2018]
Equivalent circuit model	Instinctive to be implemented and simple	Medium accuracy	[Chen, 2006], [Li, 2014]. [Plett, 2015], [Westerhoff, 2016]. Saldana, 2019]
Electrochemical	Very high accuracy	Complex and	[Ramadesigan,

model		time consuming	2012] [Moura, 2013], [Han, 2015],
Reduced order model	Less computational cost	Not an accurate model because of assumptions	[Smith, 2007],[Fan, 2018], [Domenico, 2008]
Data-driven model	Less time consumption and high accuracy	High complexity	[Meng, 2018], [Yang, 2019], [Vidal, 2020], [Liu, 2021]

Consider the following to increase performance accuracy.

- The relationship between battery variables and SOC, temperature, and current is modelled.
- Modelling of hysteresis and OCV as a function of temperature and SOC
- Aging effects are considered in the model
- RC networks are modelled at different SOC and temperatures

2.3 EIS test

The EIS test is a non-invasive, highly effective method for determining the battery's impedance properties. Impedance studies are used to analyse the battery system variables in relation to the SOC, system temperature, and current. During the experiment, a small amplitude sinusoidal AC signal is supplied to determine the system voltage, current for the specified phase, and input amplitude. Up until the battery is completely depleted, the test is repeated for various frequencies.

$$\Delta V = V_{amp} sin\ (2\pi f t) \tag{2.1}$$

$$\Delta I = I_{amp} sin\ (2\pi f t - \theta) \tag{2.2}$$

$$Z = \frac{\Delta V}{\Delta I}\ (e^{j\theta}) \tag{2.3}$$

$$Z (\omega) = |Z| (Cos (\theta) - j Sin (\theta) \tag{2.4}$$

$$Z_{re} = \acute{Z} = |Z| Cos (\theta) \tag{2.5}$$

$$Z_{Im} = Z" = |Z| Sin (\theta) \tag{2.6}$$

Where,

θ - phase shift between voltage and current

The system impedance is calculated from Eqns. (2.1) to (2.4) with regard to phase angle and magnitude. The real and fictitious elements of the system impedance are depicted in equations (2.5) and (2.6). The 18650 Li-ion cells were put to the test in order to collect the impedance data; table 2.6 provides cell specifics.

Table 2.6 Li-ion cell specifications

Lower threshold voltage	2.5V
Nominal voltage	3.6V
Upper threshold voltage	4.2V
Rated capacity	3.350Ah

The variables (SOC, ambient temperature, and SOH) that significantly impact the battery performance and characteristics are considered for this test. The operating range of the battery variables is shown in Table 2.7.

Table 2.7 EIS test battery variables

Battery Variable	Range
State of health	100%
Temperature	25°C
State of charge	100 to 0%

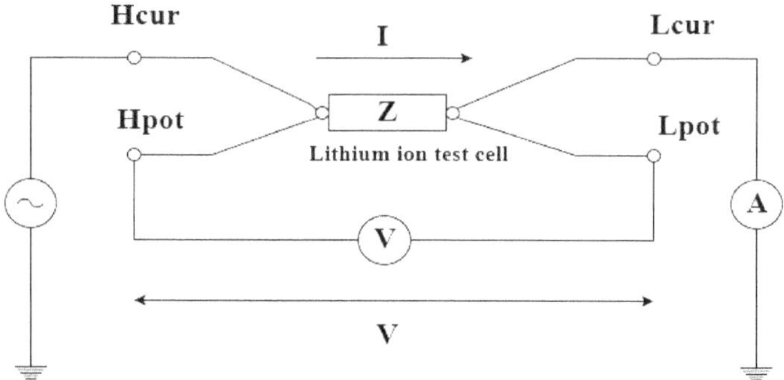

Figure 2.8 EIS test setup [Zia, 2016]

Fig. 2.8 depicts the EIS test configuration. Two reference electrodes, a working electrode, and a counter electrode make up the four probes that are connected to the cell. On the cell's negative side, a counter electrode and one reference electrode are joined. The positive side of the battery cell is connected to the working electrode and the reference electrode. The voltage sensor measures the voltage across Lpot and Hpot, whereas the current sensor measures the current flowing via Lcur and Hcur terminals.

When the cell is initially completely charged (100%), the EIS measurement is carried out by applying 3mV between the negative and positive electrode sides. For each 10% of the cell's discharge, the discharge is further paused, and the EIS measurement is then initiated. Up until the cell is completely discharged, the same process is repeated. A logarithmic scale is used to measure the impedance spectra between the frequency ranges of 10 mHz and 100 kHz. Seventy-one frequencies logarithmically spaced were measured for seven decades. The lower number of points/decade reduces the duration of the experiment, but more points allow for more precise results. 100 -10 kHz,

10 - 1 kHz

1- 100 Hz

100 - 10 Hz

10 - 1 Hz

1 Hz - 0.1 Hz

0.1 Hz - 0.010 Hz

From the impedance data, the system parameters are measured. The impedance spectrum (Nyquist plots) is drawn from 0% to 100% SOC using EIS impedance data.

2.3.1 Nyquist Plot (R, C, and L behaviour)

The Nyquist plot is used to display the impedance data obtained from the EIS test. The Nyquist plot of 0% to 100% SOC curves is displayed in Fig. 2.9 (a-k). The impedance behaviours of the battery cells taken into account at various SOCs for the analysis are displayed in Nyquist plots.

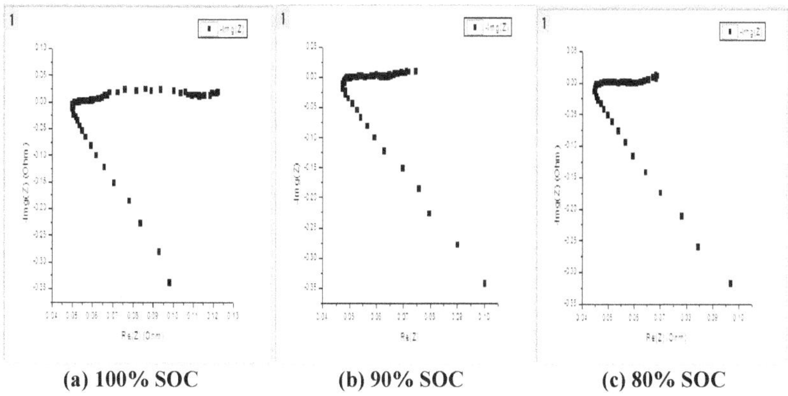

(a) 100% SOC　　　　**(b) 90% SOC**　　　　**(c) 80% SOC**

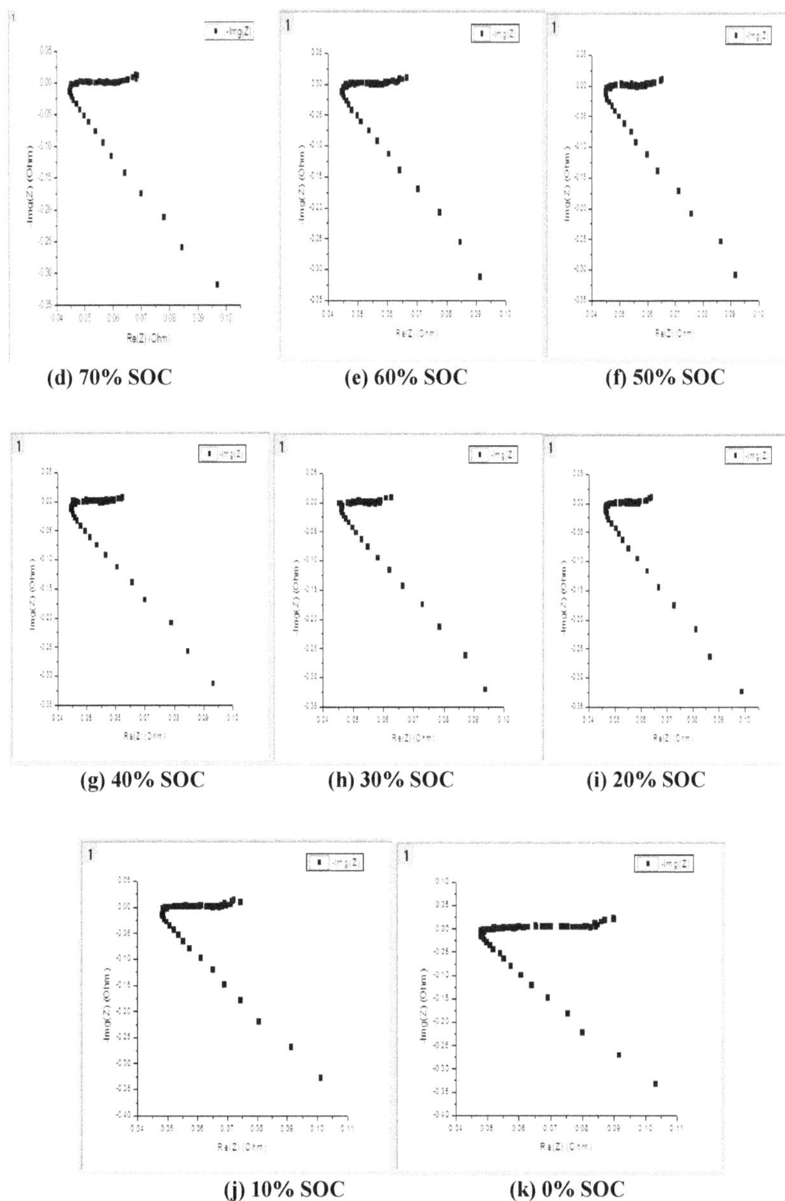

(d) 70% SOC   **(e) 60% SOC**   **(f) 50% SOC**

(g) 40% SOC   **(h) 30% SOC**   **(i) 20% SOC**

(j) 10% SOC   **(k) 0% SOC**

Figure 2.9 (a-k) Nyquist plot at different SOC considering R, L, and C

It represents the real (resistance) and imaginary (inductance and capacitance) impedance on the X and Y-axis. Battery capacitance is represented by the positive number on the Y-axis, while inductance is represented by the negative value on the impedance curve.

2.3.2 Nyquist Plot (R and C behavior)

Fig. 2.10 (a-k) displays the Nyquist plots, which solely take the battery's resistance and capacitance characteristics into account. The inductance of cable wires and connecting lines is a component of the inductive part. It is not taken into account for the analysis because its impact on the battery modelling is negligible.

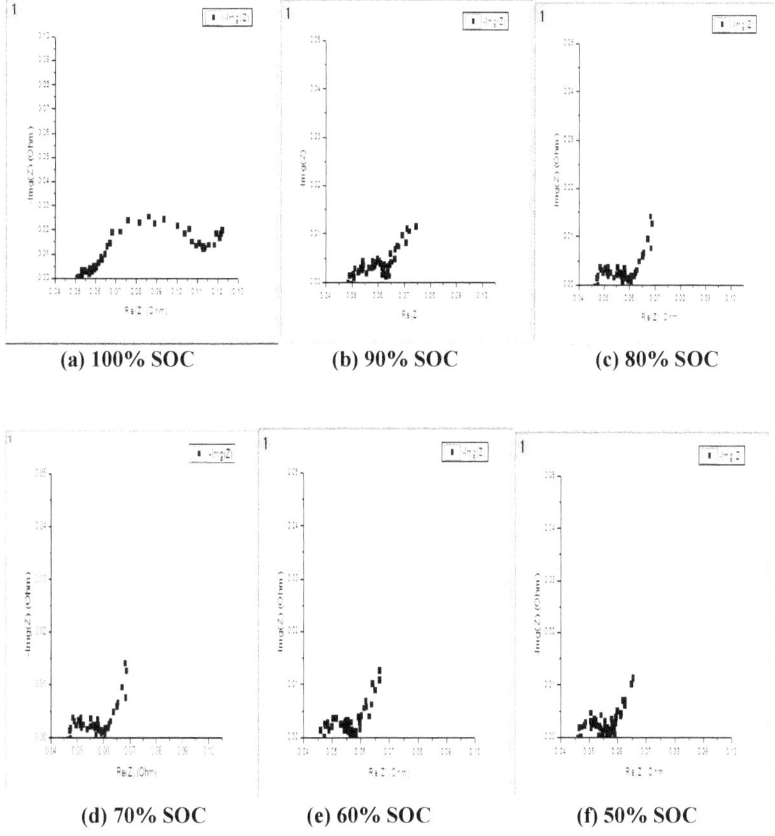

<table>
<tr><td>(a) 100% SOC</td><td>(b) 90% SOC</td><td>(c) 80% SOC</td></tr>
<tr><td>(d) 70% SOC</td><td>(e) 60% SOC</td><td>(f) 50% SOC</td></tr>
</table>

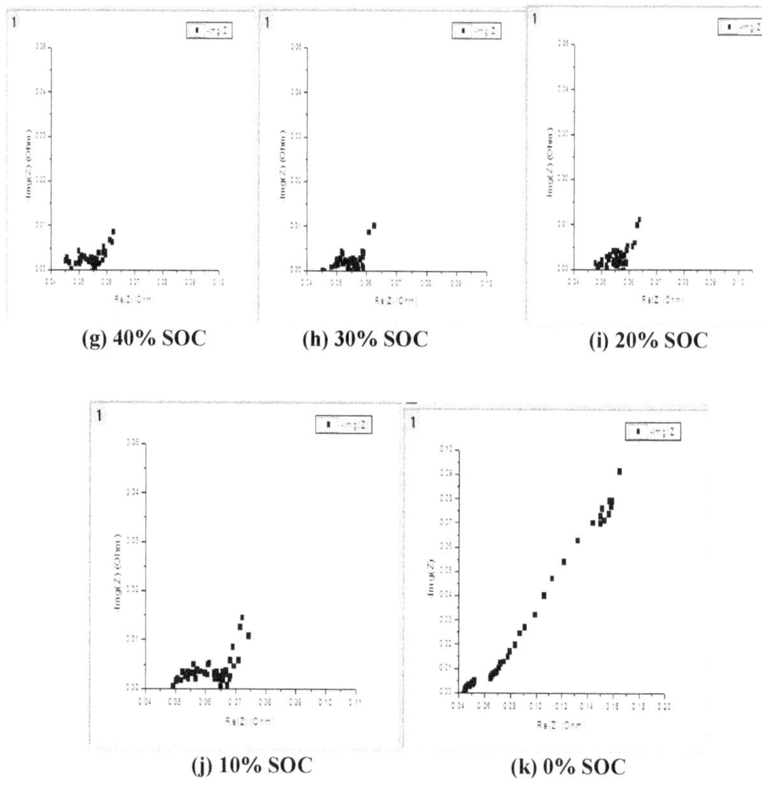

(g) 40% SOC   (h) 30% SOC   (i) 20% SOC

(j) 10% SOC   (k) 0% SOC

Figure 2.10 (a-k) Nyquist plot at different SOC considering R and C

Figure 2.11 (a) and (b) depict the high, low, and medium frequency regions of the Nyquist spectrum (b). The inductance and ohmic resistance of the battery are represented at the very high frequency area (orange colour). Due to the charge transfer effect at the electrode, the parallel RC network in the ECM marks the medium frequency area (purple hue). The diffusion phenomenon in the electrode material is shown in the low frequency area (which is shaded in green).

36

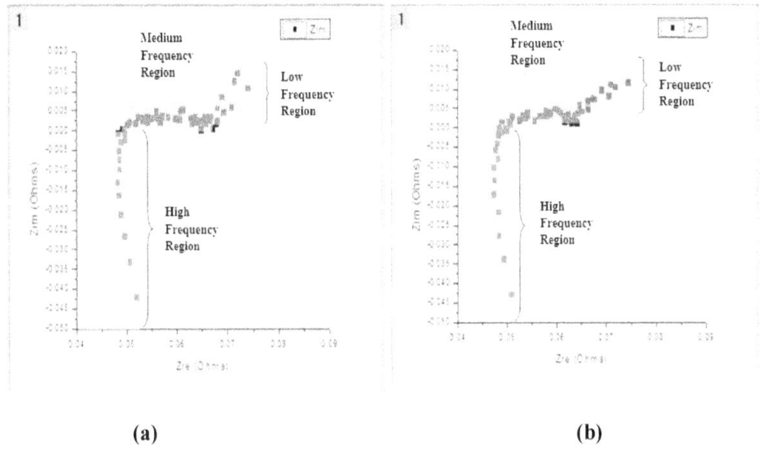

(a)   (b)

Figure 2.11 Nyquist plot (RLC) (a) at 90% SOC (b) at 10% SOC

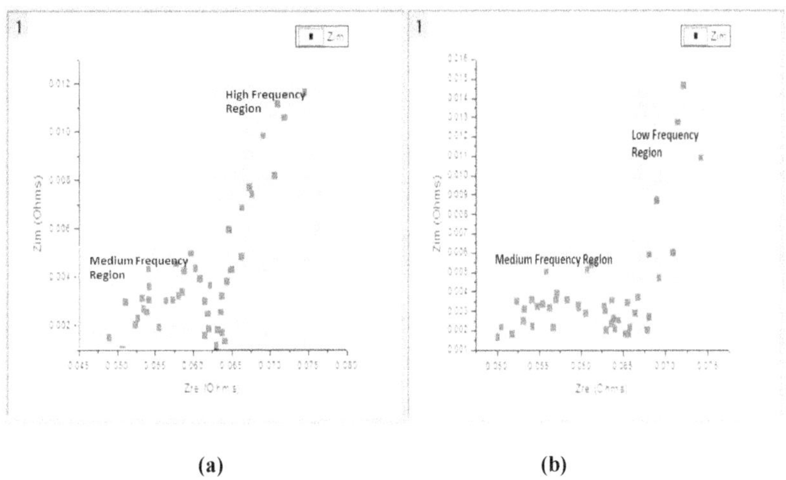

(a)   (b)

Figure 2.12 Nyquist plot (RC) (a) at 90% SOC (b) at 10% SOC

Fig. 2.12 (a) and (b) show the impedance spectra neglecting the inductive effect. Since EV batteries are generally utilized in the SOC region between 20% - 80%, the corresponding spectra are shown in Fig.2.13, and 2.14. Each point in the plot

37

shows the impedance of the battery at one frequency.

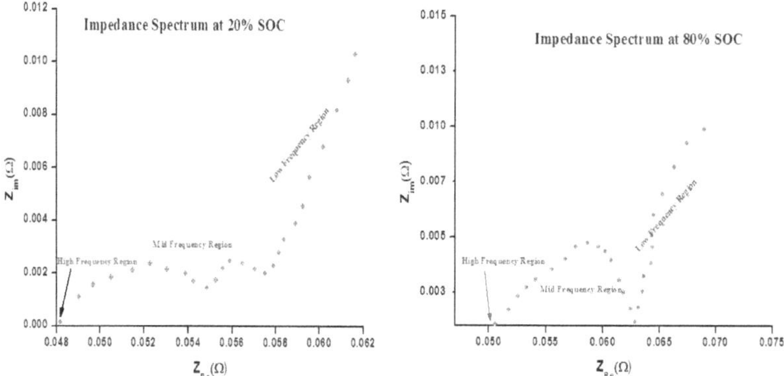

Figure 2.13 Nyquist plot at 20% SOC   **Figure 2.14 Nyquist plot at 80% SOC**

> At very high frequencies (1kHz - 100 kHz), the spectrum shows the inductive (L) behaviour and the intersection of ohmic resistance (R0) with the real axis.

> The partially overlapped semi-circle in the medium frequency (1Hz – 1 kHz) region represents the charge transfer effect at the electrodes indicated by a parallel RC pair.

> The sloping line represents the diffusion processes (Zw) in the active element of the electrodes at very low frequencies (0.01Hz - 1Hz).

From the Nyquist plot, the battery impedance varies based on the SOC. The electrode reaction dominates the spectrum at high and mid frequencies, whereas diffusion and migration dominate at low frequencies. The effect of double layer capacitance, charge transfer resistance, and Solid Electrolyte Interface (SEI) are detectable on lower SOC. From the impedance test data, the SOH, battery ageing, and internal fault of the battery can be predicted [Zhang, 2020].

38

2.3.3 Parameter estimation

From the RC ECM, the impedances of the battery can be described as follows [Zhang, 2016].

$$Z_{EC,iRC}(R,C,f) = R_0 + \sum_{i=1}^{3} \frac{R_i}{1+j\omega \cdot R_i C_i} \tag{2.7}$$

$$Z_{1RC} = R_0 + \frac{R_1}{1+j\omega R_1 C_1} \tag{2.8}$$

$$Z_{2RC} = R_0 + \frac{R_1}{1+j\omega R_1 C_1} + \frac{R_2}{1+j\omega R_2 C_2} \tag{2.9}$$

$$Z_{3RC} = R_0 + \frac{R_1}{1+j\omega R_1 C_1} + \frac{R_2}{1+j\omega R_2 C_2} + \frac{R_3}{1+j\omega R_3 C_3} \tag{2.10}$$

$$Z_{1RC} = R_0 + \frac{R_1}{1+(\omega R_1 C_1)2} - j\left(\frac{\omega C_1 R_1^2}{1+(\omega R_1 C_1)2}\right) \tag{2.11}$$

$$Z_{2RC} = R_0 + \frac{R_1}{1+(\omega R_1 C_1)2} + \frac{R_2}{1+(\omega R_2 C_2)2} - j\left(\frac{\omega C_1 R_1^2}{1+(\omega R_1 C_1)2} + \frac{\omega C_2 R_2^2}{1+(\omega R_2 C_2)2}\right) \tag{2.12}$$

$$Z_{3RC} = R_0 + \frac{R_1}{1+(\omega R_1 C_1)2} + \frac{R_2}{1+(\omega R_2 C_2)2} + \frac{R_3}{1+(\omega R_3 C_3)2} - \left(\frac{\omega C_1 R_1^2}{1+(\omega R_1 C_1)2} + \frac{\omega C_2 R_2^2}{1+(\omega R_2 C_2)2} + \frac{\omega C_3 R_3^2}{1+(\omega R_3 C_3)2}\right) \tag{2.13}$$

Eqn is a representation of the ECM impedance equation up to the third order RC model (2.7). The three RC models' impedance is displayed in equations (2.8) through (2.13). Eqn. (2.13) indicates that seven battery parameters need to be identified. The 3RC model impedance can be described as follows.

$$Z_1 = \frac{v_1(t)}{i_1(t)} = \frac{V_1}{I_1}(\cos\varphi_1 + j\sin\varphi_1) \tag{2.14}$$

$$Z_2 = \frac{v_2(t)}{i_2(t)} = \frac{V_2}{I_2}(\cos\varphi_2 + j\sin\varphi_2) \tag{2.15}$$

$$Z_3 = \frac{v_3(t)}{i_3(t)} = \frac{V_3}{I_3}(\cos\varphi_3 + j\sin\varphi_3) \tag{2.16}$$

$$Z_4 = \frac{v_4(t)}{i_4(t)} = \frac{V_4}{I_4}(\cos\varphi_4 + j\sin\varphi_4) \tag{2.17}$$

From Eqn. (2.14-2.17), v_1, v_2, v_3, v_4 are the sinusoidal AC voltage and $i_1(t), i_2(t), i_3(t),$ and $i_4(t)$ are the current flow through the battery. $\varphi_1, \varphi_2, \varphi_3,$ and φ_4 depict the angle difference between the battery voltage and current. According to Eqn. (2.18) and (2.19) through simulation, the model's structure and parameters are identified, and the equivalent impedance is calculated [Vladikova, 2004].

$$\hat{p}_M = Param.\,identify\,[\,\omega_i, Re_i, Im_i \mid S_M], \quad i = 1, 2 \ldots N \tag{2.18}$$

$$\hat{z}(i\omega) = Simulat.\,[\omega_i \mid S_M, \hat{p}_M] \tag{2.19}$$

Where,

$\quad\quad S_M$ – Structure matrix (model structure)

$\quad\quad \hat{p}_M$ – Parameters vector

The selection principle is based on the distance (Δ_i) between the estimated model and initial data. The best fit battery model gives a minimum distance, as shown in Eqn. (2.20) and (2.21).

$$\Delta_i(i\omega) = \phi_i\,(\,\hat{z}_i - Z_i\,) \tag{2.20}$$

Where,

$\quad\quad \phi_i$ – Weight of a function

$\hat{z}_i$ – Impedance obtained through simulation

$$\sum \Delta_i{}^2 \rightarrow \min \tag{2.21}$$

Using the least-square algorithm (Eqn. (2.22)), the measured impedance spectrum parameters are fitted. By using this method, the parameter values are adjusted. As a result, the residuals are minimized [Westerhoff, 2016].

$$\min_x f(x) = \min_{R,C} \sum_{n=f_{min}}^{f_{max}} [Z_{EC}\,(R,C,f_n) - Z_{measure,n}\,]^2 \tag{2.22}$$

Where Z_{EC} is a battery model complex impedance. Based on the relationship between the battery model parameters and SOC, it uses the polynomial and exponential functions.

$$f(x) = b_0 e^{b_1 x} + a_0 + a_1 x + a_2 x^2 + \dots\dots + a_m x^m \tag{2.23}$$

The parameter values are acquired by finding the minimum value of the Eqn. (2.24),

$$E(b_0, b_1, a_0, a_1, a_2, \dots. a_m) = \sum_{k=1}^{n} (b_0 e^{b_1 x_k} + a_0 + a_1 x_k + a_2 x_k^2 + \dots + a_m x_k^m - z_k) \tag{2.24}$$

Once the parameter values are derived, the relationship between the OCV, R_0, and RC parameters with respect to SOC are plotted as represented in Fig. (2.15 - 2.17)

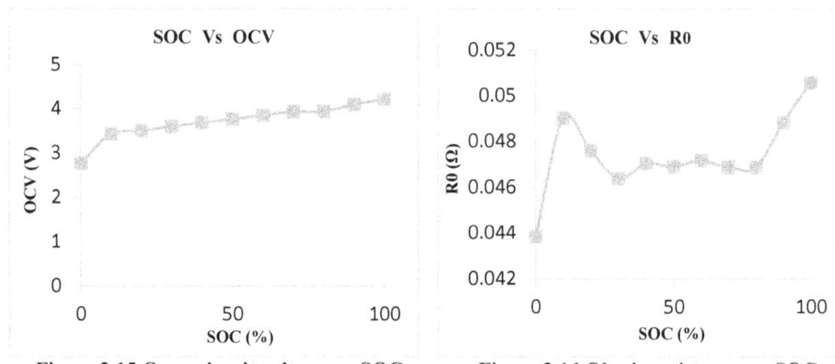

Figure 2.15 Open circuit voltage vs. SOC   Figure 2.16 Ohmic resistance vs. SOC

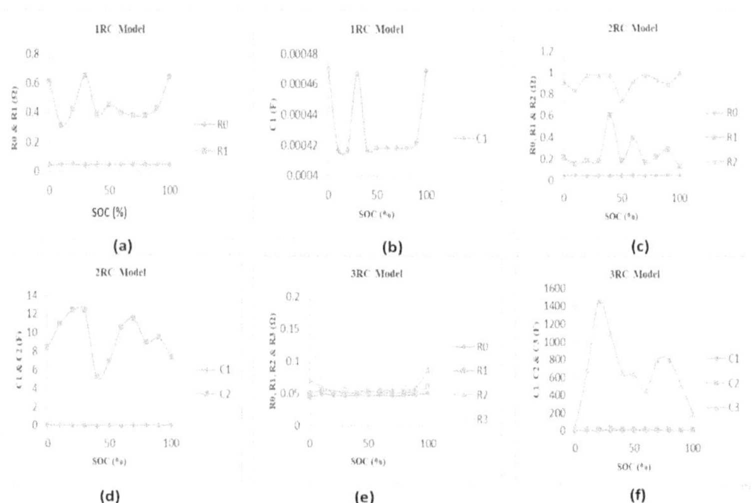

Figure 2.17 RC parameter values based on SOC

OCV is the key parameter that needs to be determined, its value is a function of temperature, and SOC is represented in Fig.2.15. The variation of R_0 with respect to SOC significantly impacts the battery voltage behavior, shown in Fig. 2.16. The higher variation of R_0 at a low and high SOC region is the key factor to the battery voltage drop. Fig. 2.17 represents the parallel RC parameter values from the low to

42

high SOC range. From the resultant graph, the Li-ion battery has a steady charge and discharge phase within the SOC range of 20% - 80%.

2.4 Battery Model (RC) Simulation and Model Validation

The RC model parameter values are fetched from the lookup table based on the battery SOC and temperature for simulation. The three RC circuit model is implemented in MATLAB. The battery is completely charged, and after that, it is completely discharged from 100% to 0% at a constant current of 0.335 A. The time taken for full discharge is 35500 sec. The simulated RC model voltages are compared against experimental data, and error voltages are calculated to validate the model accuracy, as represented in Fig. 2.18.

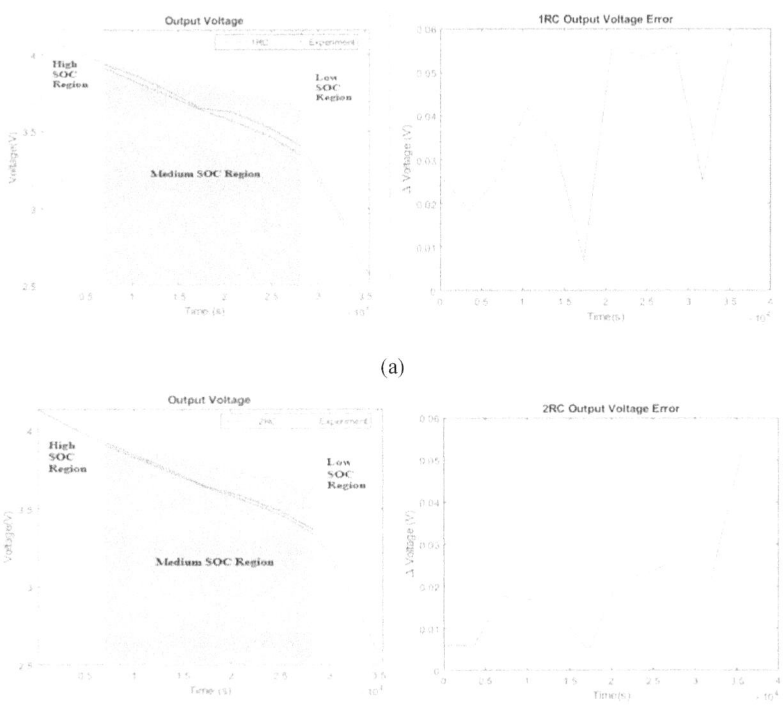

(a)

(b)

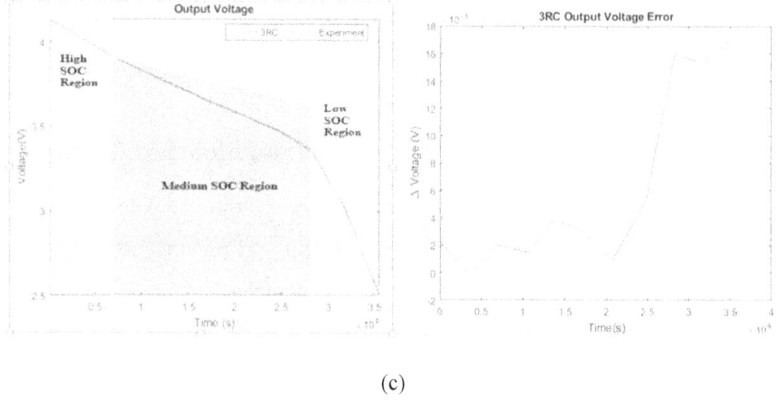

(c)

Figure 2.18 RC Model voltage comparison between hardware and simulation (a) 1RC, (b) 2RC, and (c) 3RC

The voltage deviation between 1RC and experimental voltage is high because the diffusion behaviour of the battery is not included in this model. The battery voltage shows a strong non-linear decrease at the discharge end due to the impedance rise at low SOC. From Fig. 2.18 (a), it is observed that the error voltage of the1RC model rises to 0.06V. The error voltage of 2RC is minimized due to the consideration of diffusion characteristics and the voltage error is approximately 0.05 V, as represented in Fig. 2.18 (b).

The voltage difference is minimum in the 3RC model, and the maximum error voltage is less than 0.018V, as shown in Fig. 2.18 (c). From the model voltage waveforms, the voltage difference is maximum at the final stage of the discharge due to the polarization process and electrochemical reaction happening inside the battery. Due to that, the internal resistance of the battery increases.

44

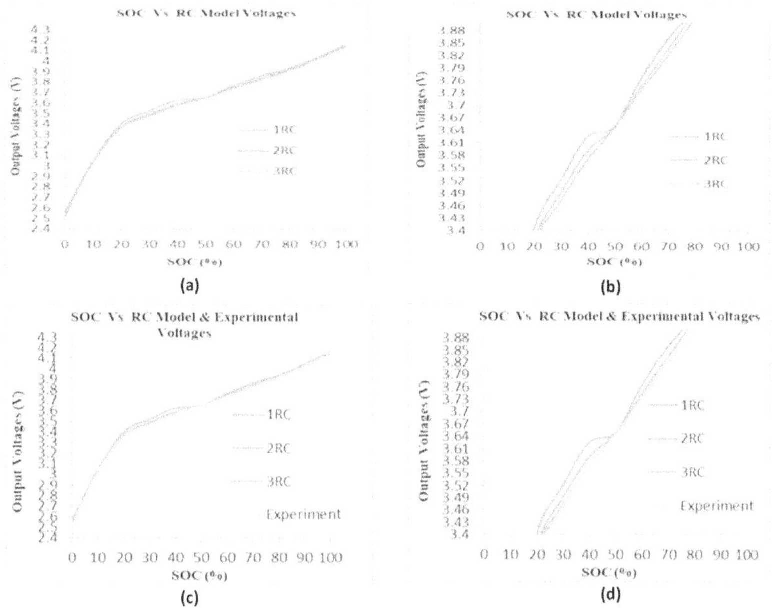

Figure 2.19 (a-d) Comparison of experimental voltage with the model voltage

In Fig. 2.19(a-d), the experimental voltage with regard to SOC is contrasted with the RC model voltages. Given that it takes the battery's diffusion and charge transfer properties into account, the 3RC voltage differential is at its lowest in the high and medium SOC zone. Due to the non-linear properties of the battery, the voltage differential is large at the lower SOC. It is an appropriate model for EV batteries because the balancing is enabled at the conclusion of charging.. The literature study concludes that, the type of RC model selection is based on the battery chemistries. In this research, all the RC models are compared through the experimental data and found that, the 3RC model is suitable for this research work. Maximum accuracy can be obtained by adding the number of RC elements. At the same time, it increases the system's complexity. Hence, the 3RC could be an optimal choice for EV applications.

45

2.5 Conclusion

One of the key procedures for obtaining precise battery condition estimation is battery modelling. Depending on the type of application, many internal and external factors may influence the battery SOC and the SOC calculation approaches. The 3RC ECM captures the dynamic characteristics of the battery better than lower order models, and it is considered for the cell balancing analysis. The selected 3RC battery model and its parameter values obtained through EIS test measurement are further used for the balancing system implementation in the succeeding chapters.

CHAPTER 3

Passive Cell Balancing in Electric Vehicle Battery Management System

3.1 Introduction

This chapter describes the implementation of a passive cell balancing system for NMC cells that covers the maximum and minimum degrees of imbalance. It also evaluates the balancing current, balancing time, energy loss, efficiency, and cost. When the EV is turned off and the battery is being charged, the balancing is done. To simulate the model in Matlab / Simscape, the estimated model parameters from the impedance test were fetched into the lookup table. To calculate the balancing current, balancing time, power loss, and system efficiency, the battery's SOC, OCV, terminal voltage, temperature, and current are used. The results of simulations aid in further designing, validating, and fine-tuning the hardware.

3.2 Cell balancing

3.2.1 Causes and impact of cell imbalance

One of the most important characteristics of BMS is cell balancing, which encourages evenly spreading the charge throughout the battery cells to ensure optimal performance. In Fig.3.1, the underlying causes of the cell imbalance are depicted.

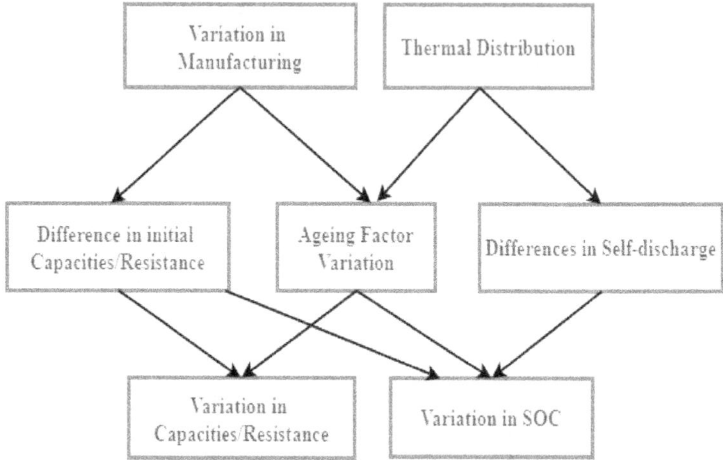

Figure 3.1 Causes for the cell imbalance [Lelie, 2018]

Overcharged or undercharged batteries can permanently harm the battery pack or speed up cell ageing. In EVs, the cell imbalance that occurs during battery charging and discharge is a serious problem. Battery pack deterioration results from the battery pack's capacity declining quickly. If the pack contains numerous cells connected in a series and is often charged, the issue gets worse. One consequence of cell imbalance is

- Early charge and discharge termination,

- safety risks from overcharge,

- premature cell degradation caused by overvoltage, and more

As seen in Fig.3.2, the capacity of the battery pack is lowered if any one of the cells in the pack exceeds the charge voltage threshold (4.2V). The capacity of the remaining healthy reservoir is not used if any of the weak cells fall below the discharge voltage threshold (2.5V), as shown in Fig. 3.3. In order to resolve these problems, a suitable cell balancing mechanism is needed.

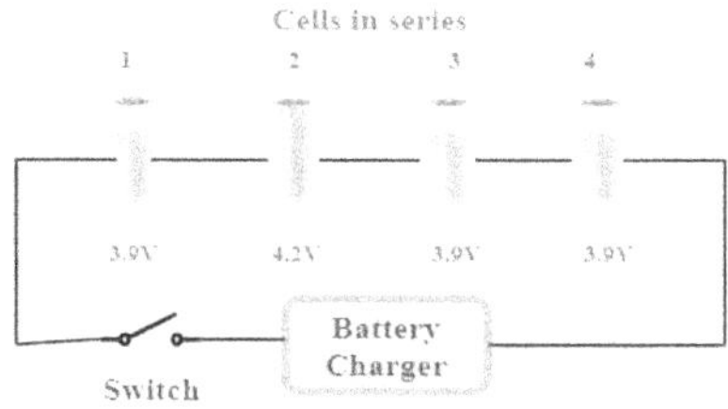

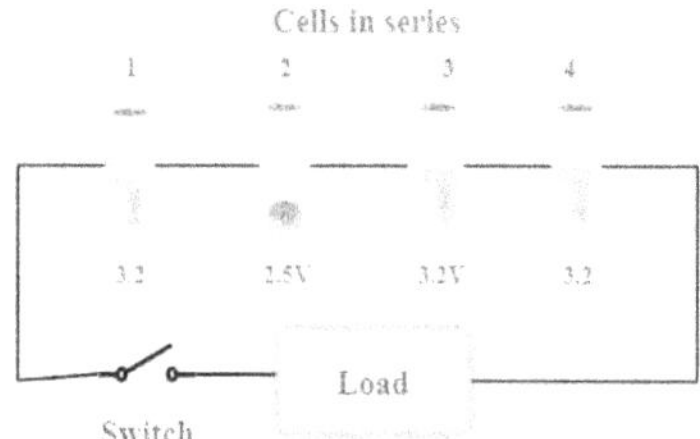

Figure 3.2 Early charge termination   **Figure 3.3 Early discharge termination**

Cell balancing circumvent the cell imbalance to improve battery life and safety. The balancing algorithms, which are divided into passive and active cell balancing, detect and correct cell imbalances [Amin, 2017].

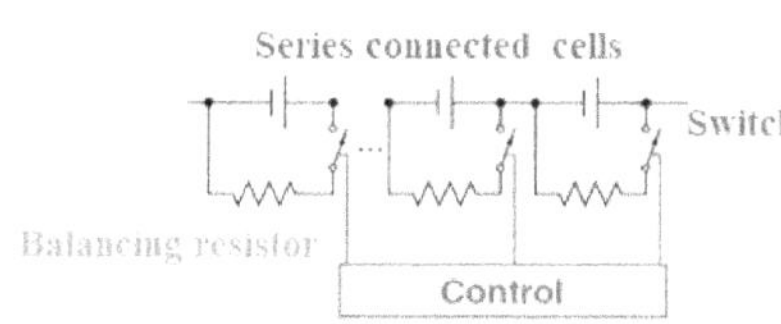

Figure 3.4 Passive cell balancing   **Figure 3.5 Active cell balancing**

As shown in Fig.3.4, in a passive system, the balancing resistor reduces the high voltage cell's excess energy until the voltages of the other cells are equal. According to Fig. 3.5, the active system in the battery pack transfers energy from high voltage cells to low voltage cells. As shown in Fig. 3.6, the cell balancing system utilised in EVs normally has three modes of operation.

49

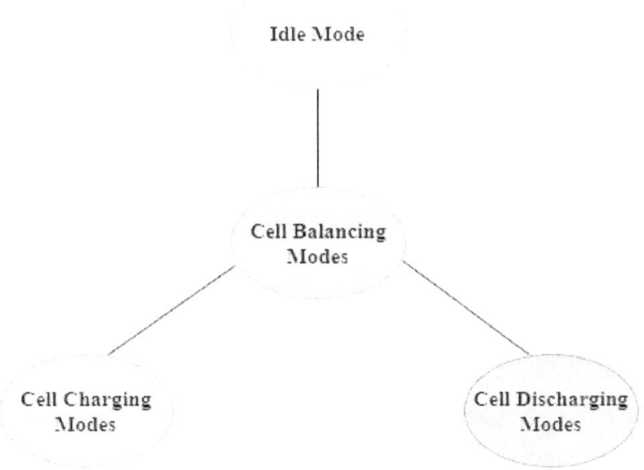

Fig. 3.6 Cell balancing modes [Omariba, 2019]

Balancing is applied to get the most out of the energy delivered to or from the cell during the battery charging or discharging. The cell must not be charged or drained in order for the idle balancing mode to function. Passive balancing is suitable for charging and not for discharging as some of the battery energy is dissipated as heat (Koseoglou, 2020). The active balancing system works in all three modes [Omariba, 2019].

3.2.2 Balancing architecture in BMS

Fig.3.7 depicts the cell balancing design of the BMS. The battery cell controller's analogue front end measures the pack voltage, cell voltage, and pack current. A serial peripheral interface is used to transmit the data to the pack controller (SPI). Utilizing temperature sensors, the microprocessor determines the battery's temperature. Constant current and voltage (CC-CV) charging for the pack is monitored and managed by the pack controller. Through standardised automotive protocols like the Controller Area Network (CAN) bus, the microcontroller communicates with the other system components.

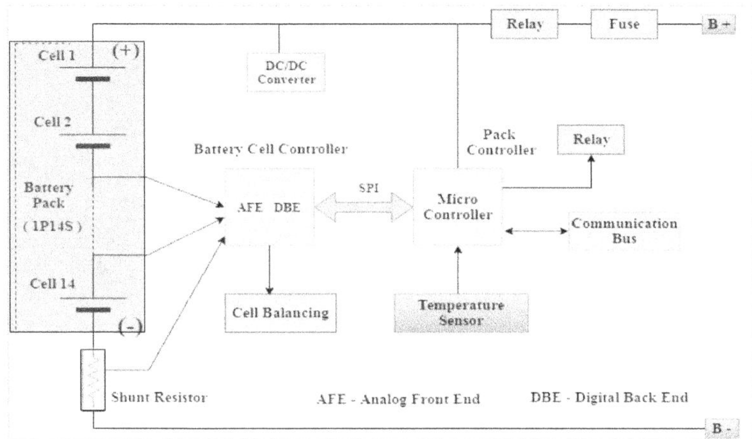

Figure 3.7 Cell balancing architecture in BMS

The battery pack controller checks for anomalies in the cell voltage when it is attached to the charger. The cell-balancing module is then turned on by the pack controller via the battery cell controller. While the cells are balancing, the charger supplies the current until all of the cells have a voltage of 4.2V. When the pack's charge hits the top threshold, the pack controller shuts off the charger.

3.2.3 Battery pack architecture

18650 Li-ion cells in an 18P14S combination with a 60Ah capacity are utilised in the battery pack for the passive balancing system implementation to produce the 48 V needed for the electric bike. The cell's respective threshold voltages for charging, discharging, and cell balancing are 4.2V, 2.5V, and 3.9V.

3.3 Passive system simulation and prototyping

3.3.1 3RC Battery model

The battery model used to help formulate the ideal cell balancing problem is the Thevenin 3RC electrical ECM. The EIS test in Chapter 2 is used to acquire the model parameters. The ECM with three parallel RC pairs is depicted in Fig. 3.8.

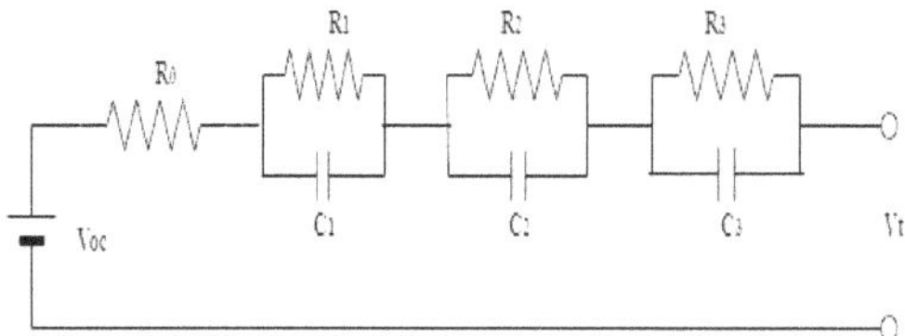

Figure 3.8 3RC model

$$V_t = V_{oc} - R_0 i - V_{R_1C_1} - V_{R_2C_2} - V_{R_3C_3} \qquad (3.1)$$

$V_{R_1C_1}, V_{R_2C_2}, V_{R_3C_3}$ - voltage of the shunt RC pair

Eqn. (3.1) represents the output voltage of the 3RC model, and Eqn. (3.2- 3.4) represent the transient voltage response of the system.

$$\frac{dV_{R_1C_1}}{dt} = \frac{i}{C1} - \frac{V_{R_1C_1}}{R_1C_1} \qquad (3.2)$$

$$\frac{dV_{R_2C_2}}{dt} = \frac{i}{C2} - \frac{V_{R_2C_2}}{R_2C_2} \qquad (3.3)$$

$$\frac{dV_{R_3C_3}}{dt} = \frac{i}{C3} - \frac{V_{R_3C_3}}{R_3C_3} \qquad (3.4)$$

$\frac{dV_{R_1C_1}}{dt}, \frac{dV_{R_2C_2}}{dt},$ and $\frac{dV_{R_3C_3}}{dt}$ – indicate transient voltage change

The first, second, and third parallel RC networks' respective time constants are denoted by R 1 C 1, R 2 C 2, and R 3 C 3. The discrete version of the equations created using ECM is represented as Equations (3.5) to (3.8) [How, 2019].

$$V_{1,j+1} = V_{1,j}e^{(-\Delta T/\tau_1)} + [1 - e^{(-\Delta T/\tau_1)}R_1I_j] \qquad (3.5)$$

$$V_{2,j+1} = V_{2,j}e^{(-\Delta T/\tau_2)} + [1 - e^{(-\Delta T/\tau_2)}R_2I_j] \qquad (3.6)$$

$$V_{3,j+1} = V_{3,j}e^{(-\Delta T/\tau_3)} + [1 - e^{(-\Delta T/\tau_3)}R_3I_j] \qquad (3.7)$$

$$V_{t,j+1} = V_{oc,j+1} + R_0 I_{k+1}(t) + V_{1,j+1} + V_{2,j+1} + V_{3,j+1}$$

$$, j \in \{1, 2, ... N\} \tag{3.8}$$

Where,

τ_1, τ_2 and τ_3 , indicate the time constant of the 3RC network.

The battery terminal voltages $V_t, V_{oc,}, and\ V_1, V_2, V_3$ represent the OCV and voltage drop across the 3RC branch's polarization resistance, respectively. Eqn is a representation of the formula to determine the battery's SOC (3.9).

$$SOC = SOC_0 - \frac{\eta}{C} \int I(t)dt \tag{3.9}$$

C – Capacity of the battery, SOC_0 -Initial SOC of the battery, η -   Coulombic efficiency. The current integration (Coulomb counting) method is used to determine the battery SOC. Since the self-discharge rate of Li-ion batteries is minimum, the coulomb efficiency value is close to 1.

$$\eta = \frac{Capacity\ discharged}{Capacity\ charged} \tag{3.10}$$

According to Eqn. (3.10), η lowers when more battery charge and discharge cycles are performed. Battery testing under charge and discharge circumstances yields a link between battery voltage and SOC. The charging current is provided by equation (3.11).

$$I_{Ch,j}(t) = I_{B,j}(t) + I_{C,j}(t) \tag{3.11}$$

$I_{C,j}$ – Cell current, $I_{B,j}$ – Balancing current. For N cells in series, the current and voltage of the pack are shown in Eqn. (3.12) and (3.13)

$$I_{C,j}(t) = I_{C,j+1}(t), \ j \in \{1, 2, ... N\text{-}1\} \tag{3.12}$$

$$V_o(t) = \sum_{j=1}^{N} V_j(t) \tag{3.13}$$

The output of the battery system is represented in Eqn. (3.14) and (3.15),

$$I_o(t) = \sum_{j=1}^{N} (I_{Bo,j}(t) + I_{C,j}(t)) \tag{3.14}$$

$$P_o(t) = V_o(t)I_o(t) \tag{3.15}$$

I_o - output current, $I_{Bo,j}(t)$ - output current of the balancing circuit, P_o –Output power from the battery pack. The power balance equation of each cell during the balancing is shown in Eqn. (3.16),

$$V_j(t)I_{B,j}(t) = V_o(t)I_{Bo,j}(t) + P_{lB,j}(t) \tag{3.16}$$

$P_{lB,j}$ – Power loss across the balancing circuit. According to Eqn. (3.17) and Eqn.(3.18) and the related balancing current and balancing resistor, the balancing circuit's power loss is determined,

$$\text{Balancing current, } I_B = \frac{V_B}{R_B} \tag{3.17}$$

$$P_{lB,j}(t) = R_B I_{B,j}^2(t) \tag{3.18}$$

Then Eqn. (3.19) formulates the balancing circuit current constraints,

$$I_B^{min} \leq I_{B,j}(t) \leq I_B^{max} \tag{3.19}$$

I_B^{max}, I_B^{min}- the maximum and minimum balancing current limits for the circuit. In order to obtain the voltage and thermal balance during balancing, the following restrictions are introduced:

$$\left|V_j(t) - V_{avg}(t)\right| \leq \Delta_V \tag{3.20}$$

Δ_V –Maximum voltage difference between cell and pack average

$$\left|T_j(t) - T_{avg}(t)\right| \leq \Delta_T \tag{3.21}$$

Δ_T - Maximum temperature disparity between the average of each cell and the pack. The change in battery temperature is calculated using the first-order differential equation, as given in Eqn (3.22),

$$T_B(t) = T_{int} + \int_0^t \frac{\left[P_L - \frac{(T-T_a)}{R_T}\right]}{C_T} \, d\tau \tag{3.22}$$

T_a – the ambient temperature in °C, T_B – battery temperature in °C, T_{int} –battery initial temperature equal to ambient temperature in °C, C_T – thermal capacitance in Joules/°C, R_T - thermal resistance in °C /Watts, P_L – power loss across the battery internal and polarization resistance. For the 3RC equivalent battery model, the power loss is calculated by using the Eqn. (3.23),

$$P_L = I^2 R_0 + I^2 R_1 + I^2 R_2 + I^2 R_3 \tag{3.23}$$

The value of R_1, R_2, R_3, and R_0 are estimated at different SOC through EIS test measurement in the preceding chapter 2.

3.3.2 Modeling outcomes

The implemented passive balancing subsystem model in Matlab/Simscape includes voltage measurement block, parallel RC pairs, charge and discharge current measurement, current pulse generator, SOC and temperature measurement, and ambient temperature simulator as shown in Fig. 3.9. The values of the equivalent circuit elements obtained from the lookup table are the function of battery temperature and SOC in a parameter estimation procedure.

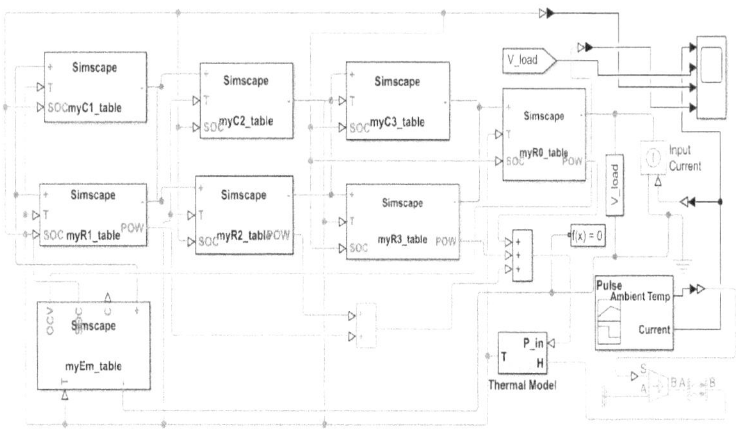

Figure 3.9 3RC Matlab-Simscape model (2019b)

With a pulse charge current of 0.7A and a duty cycle (D) of 1,620 secs "on" and 1,800 secs "off," the battery is fully charged from 0% to 100% SOC. The charge current waveform is shown in Fig. 3.10, and a full charge takes 35,500 seconds. The battery voltage is depicted under constant pulse current charge in Fig. 3.11. During the intervals of repose, the battery voltage demonstrates the dynamic response between charge current pulses.

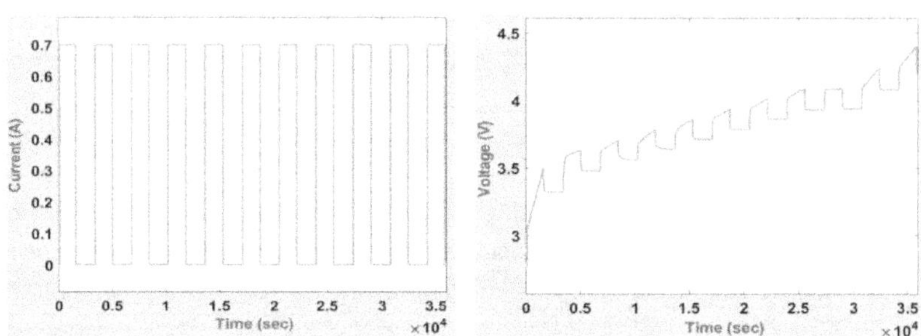

Figure 3.10 Charging current
Figure 3.11 Battery voltage (charging)

According to Fig.3.12, the SOC of the battery increased over the course of the charge period. The SOC is calculated from the battery current. Due to the internal resistance variation, it is seen that the temperature is rising from the ambient temperature of 20°C to 20.85°C at the beginning of charging and reaching 20.7°C at the end of charging. Fig. 3.13 illustrates the increased temperature cycle.

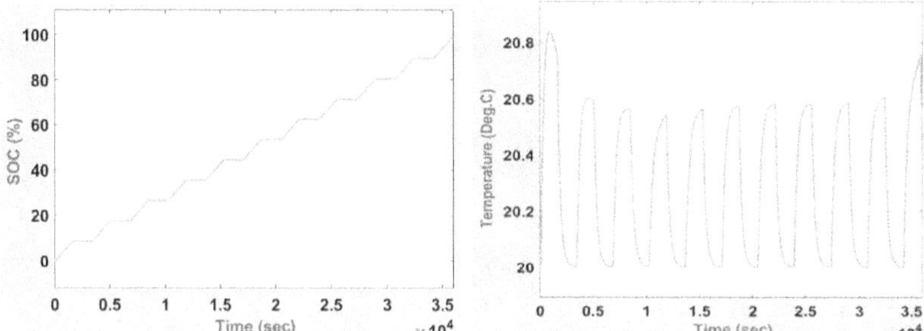

Figure 3.12 Battery SOC (charging)   **Figure 3.13 Battery temperature (charging)**

The battery was pulse discharged from its whole SOC after being fully charged. With a duty cycle of 1,620 seconds "on" and 1,800 seconds "off," the pulse current was 0.7 A. The battery discharge current waveform is shown in Fig. 3.14, and the total discharge time is 35,500 sec. The battery voltage during constant pulse current discharge is shown in Fig. 3.15, and just before the discharge is finished, there is a significant non-linear decline in the voltage. As the SOC level decreases, the cell impedance increases.

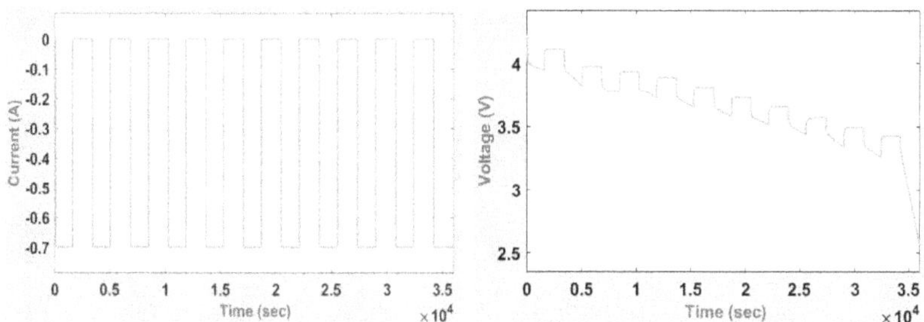

Figure 3.14 Discharging current   **Figure 3.15 Battery voltage (discharging)**

The voltage is 2.5 V at 0% SOC and 4.2 V at 100% SOC. The battery SOC's declining trend over the course of the discharge period is shown in Fig. 3.16. Due to the internal resistance change and the prolonged temperature cycle depicted in Fig. 3.17, it is seen that the temperature rises from the ambient temperature of 20°C to 20.9°C at the completion of discharge. The selection of pulse attributes (current amplitude, duty cycle etc.) influences the charging/discharging time, battery impedance parameters, and life span of the battery.

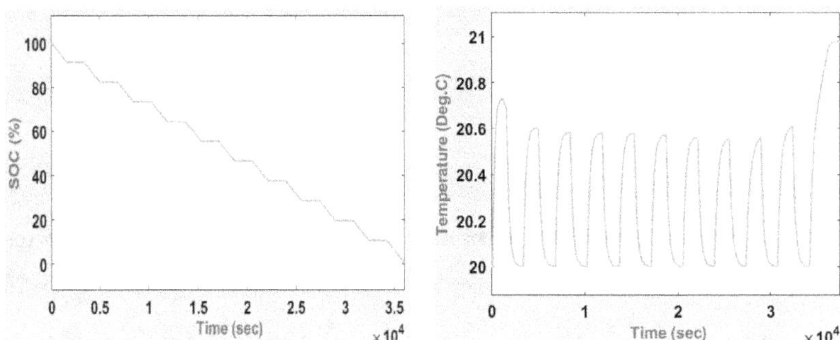

Figure 3.16 Battery SOC(discharging) Figure 3.17 Battery temperature (discharging)

The Li-ion battery exhibits complicated electrochemical characteristics while draining and charging, and the model accurately predicts the battery SOC, according to the simulation findings. The cell voltage information shown in Fig. 3.18 serves as the foundation for the proposed balancing method.

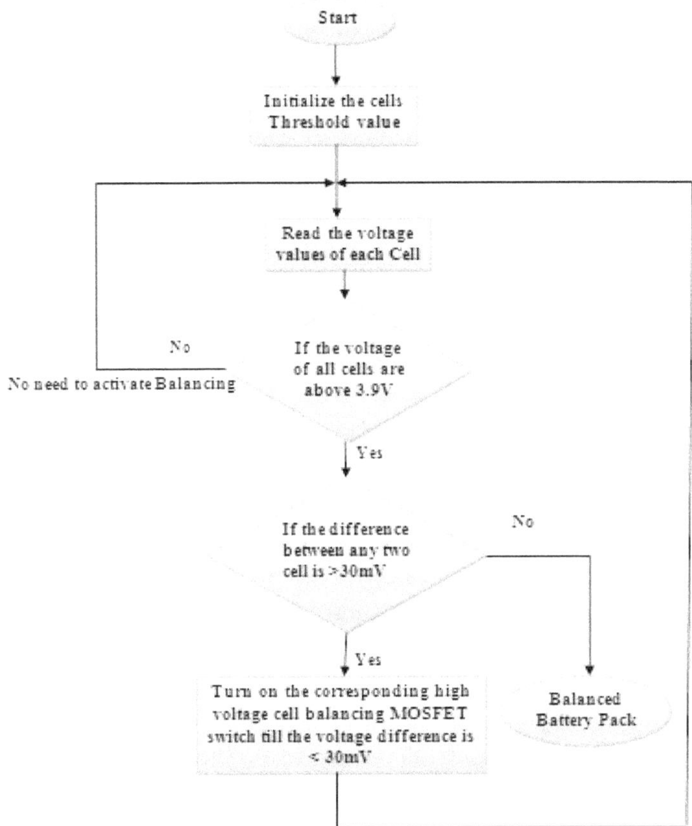

Figure 3.18 Passive system flowchart

Due to the minimum SOC variations among the cells with respect to the difference in voltage and monitoring circuitry accuracy and resolution 30mV difference has been considered in this work. When every cell in the pack reaches the set threshold, the cell balancing system is engaged, and the balancing process begins. The voltage of the cell and the value of the balancing resistor across the cell determine the balancing current. More resistors and MOSFET switches are used in the balancing circuit design in the passive system implementations, as shown in Fig. 3.19. Balancing current determines the speed of the energy transfer, and the passive system uses a low balancing current to minimize heat dissipation. The balancing current is calculated from Eqn. (3.24).

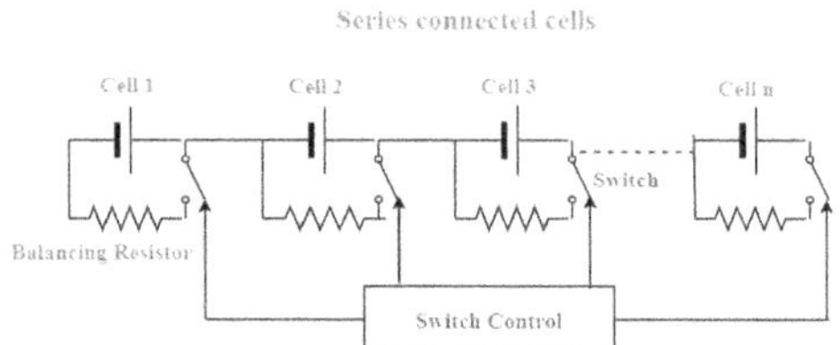

Figure 3.19 Passive cell balancing system

$$I_B = \frac{V_C}{R_B} \tag{3.24}$$

Where,

- V_C - cell voltage
- R_B - balancing resistor

Energy loss = Power loss × Balancing time $\hspace{3em}$ (3.25)

$$E_{loss} \quad = \frac{V^2}{R_B} \times t_B \quad \text{or} \quad I_B^2 R_B \times t_B \tag{3.26}$$

From Eqn. (3.25) and (3.26), the energy loss across the balancing resistor is calculated. The minimum balancing time during passive balancing leads to a higher current requirement on the MOSFET switch and a high wattage rating on the resistors.

3. 3. 3 System prototyping

The passive balancing system prototyping is designed and implemented using an NXP battery cell controller for validating the passive system's functionality and performance, as shown in Fig.3.20. The battery pack contains the balancing system. An NXP semiconductor battery cell and pack controller serves as the passive system's central controller. The individual cell and pack voltages are monitored by the MC33771 cell controller. Based on the voltage measurement, the pack controller manages the switches S1–S14.

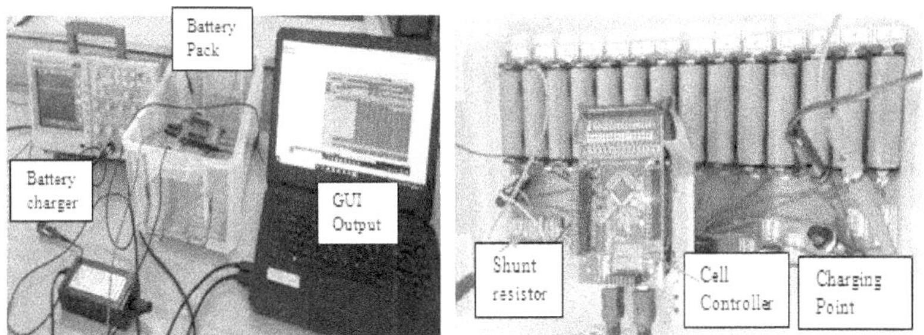

Figure 3.20 Experimental setup

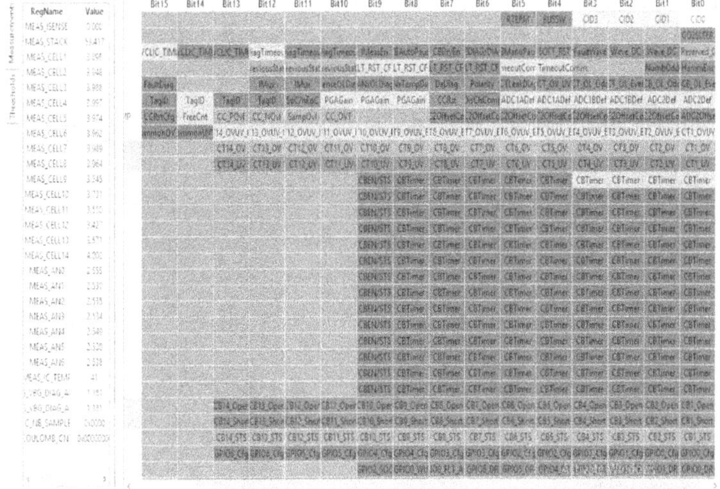

Figure 3.21 Battery parameters in GUI

GUID-CID-0B220E1308
GUID-CID-02
GUID-CID-03
GUID-CID-04
GUID-CID-05
GUID-CID-06
GUID-CID-07
GUID-CID-08
GUID-CID-09
GUID-CID-10
GUID-CID-11
GUID-CID-12
GUID-CID-13
GUID-CID-14
GUID-CID-15

Time	CID	Status	Command Result	Visense[m	Vstack[V]	Vct1[V]	Vct2[V]	Vct3[V]	Vct4[V]	Vct5[V]	Vct6[V]	Vct7[V]	Vct8[V]	Vct9[V]	Vct10[V]	Vct11[V]	Vct12[V]	Vct13[V]	Vct14[V]
582894	CID-1	0		0	0	55.432	3.995	3.934	3.998	3.997	3.984	4.005	3.999	3.974	3.955	3.939	3.96	3.935	3.982	3.931
582974	CID-1	0		0	0	55.435	3.996	3.935	3.997	3.996	3.984	4.005	3.998	3.974	3.956	3.939	3.959	3.935	3.982	3.931
583053	CID-1	0		0	0	55.432	3.994	3.934	3.997	3.996	3.984	4.005	3.998	3.973	3.955	3.939	3.96	3.936	3.982	3.931
583133	CID-1	0		0	0	55.433	3.995	3.934	3.997	3.996	3.984	4.005	3.998	3.973	3.956	3.939	3.96	3.936	3.982	3.932
583213	CID-1	0		0	0	55.433	3.995	3.934	3.998	3.996	3.983	4.005	3.998	3.973	3.955	3.939	3.96	3.935	3.982	3.931
583292	CID-1	0		0	0	55.424	3.995	3.934	3.998	3.996	3.984	4.005	3.999	3.973	3.955	3.939	3.959	3.935	3.983	3.931
583372	CID-1	0		0	0	55.429	3.994	3.933	3.997	3.996	3.984	4.005	3.998	3.974	3.955	3.939	3.96	3.935	3.982	3.931
583452	CID-1	0		0	0	55.429	3.995	3.933	3.998	3.997	3.984	4.005	3.999	3.974	3.955	3.938	3.96	3.936	3.982	3.931
583532	CID-1	0		0	0	55.439	3.994	3.934	3.998	3.996	3.984	4.005	3.999	3.974	3.955	3.939	3.96	3.935	3.982	3.931
583611	CID-1	0		0	0	55.432	3.994	3.934	3.998	3.997	3.984	4.004	3.999	3.973	3.956	3.939	3.96	3.936	3.983	3.931

Figure 3.22 GUI data logging

In Figures 3.21 and 3.22, the battery cell voltages, pack current, balancing current, pack temperature, and switch status (S1-S14) are displayed on the GUI and recorded using the data logging system.

3. 4 Results and discussion

3. 4. 1 Simulation outcome

When the cell voltage reaches 3.9 V, which equates to an approximate SOC of 80%, the balancing algorithm begins to function. The charging current is reduced at the high SOC area, and the imbalance brought on by the voltage drop across the resistor has a very small effect on cell voltages. So, towards the end of charging, the last voltage-based balancing is proposed [Andrea, 2010]. The balancing threshold voltage is set at 30 mV. There are 14 cells used in the simulation. To illustrate the balancing actions better, two of the cells are taken for the explanation.

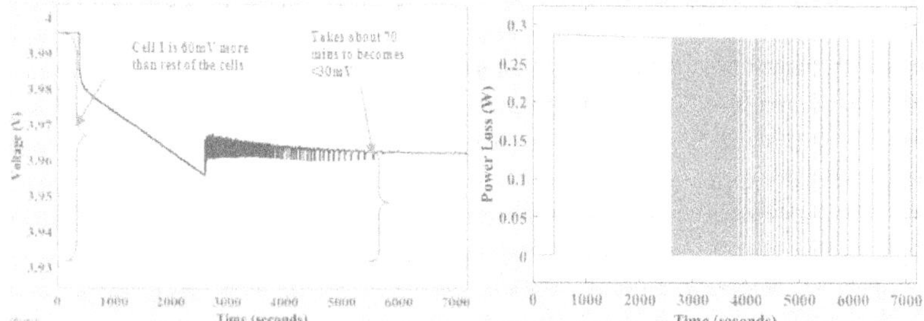

Figure 3.23 Cell1 voltage during balancing   Figure 3.24 Power loss across cell1 resistor

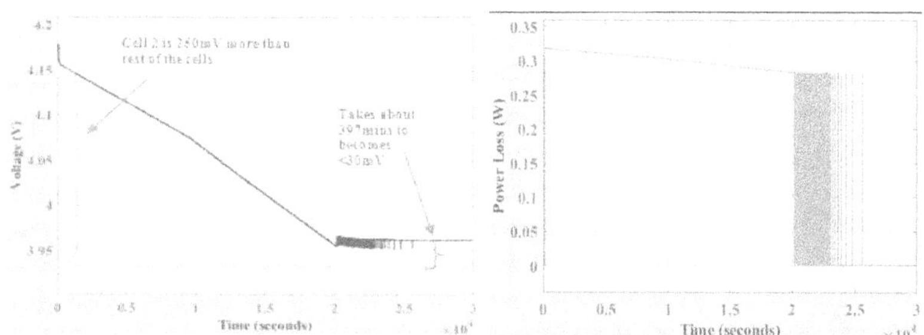

Figure 3.25 Cell2 voltage during balancing   Figure 3.26 Power loss across cell2 resistor

The voltage of cell 1 (the overcharged cell) is set to 3.994 V during simulation, while the voltages of the other cells are maintained at 3.934 V with a voltage differential of 60 mV. The voltage differential decreases to less than 30 mV after approximately 70 minutes of balancing, as shown in Fig. 3.23. Fig. 3.24 depicts the power dissipation across the shunt resistor. Fig. 3.25 shows the degree of imbalance (which can range up to 250 mV) between old and new cells and the balancing time of 397 minutes. Due to the greater voltage differential between the cells, it is noted that the initial power dissipation across the balancing resistor is significant and that it takes longer to reach the steady state, as shown in Fig. 3.26.

3. 4. 2 Hardware realization

The 14 cells have been added to the NXP controller-based hardware, and data has been acquired using the GUI interface that is already there. The cell voltage during the balancing period was recorded using a digital signal oscilloscope. The NXP hardware-based system's cell balancing process captured the necessary data.

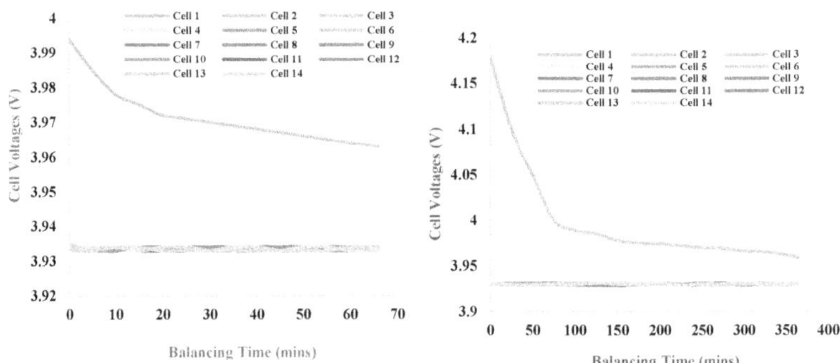

Figure 3.27 Balancing cell voltage from GUI at ΔV=60mV and 250mV

Figure 3.27 shows the balancing time in relation to cell voltage for the situations with balancing discrepancies of 60 mV and 250 mV. It has been noted that the cell voltage first drops quickly and there is non-linearity across the cell voltage, which eventually settles. Tables 3.1 and 3.2 compare and tabulate the hardware and simulation data.

3. 5 Energy efficiency, loss, and cost calculation

The passive balancing system is discussed in this section in terms of energy efficiency, energy loss, and cost analysis.

3. 5. 1 Energy loss and efficiency

The quantity of energy inserted into and expelled from the pack is used to calculate energy loss [Aizpuru, 2017]. Eqn. (3.27) and (3.28) give the charging energy and energy loss across the balancing resistor.

$$E_C = \int_0^t V_{Bat}\,(t)\,I_{Bat}\,(t)dt \qquad\qquad (3.27)$$

E_C - Total energy fed from the battery charger.

$$E_{loss} = \int_0^t P_{LBS}\ (t)dt \tag{3.28}$$

E_{loss} - Energy loss across the balancing resistor

$$E_{C-Bat} = E_C - E_{loss} = \int_0^t \sum_1^n V_{b,n}(t)\ I_{c,n}\ (t)dt \tag{3.29}$$

$$Energy\ efficiency = \frac{E_{C-Bat}}{E_C} \tag{3.30}$$

E_{C-Bat} - Total energy available in the battery pack after balancing, computed by integrating each cell's power (VI) with respect to time, is shown in Eqn. (3.29). The battery pack's energy efficiency is calculated from Eqn. (3.30).

3. 5. 2 Energy loss, efficiency, and cost analysis

Table 3.1 Energy loss, efficiency, and cost analysis of passive balancing system

ΔV(mV)	Cell s (#)	From the Simulation analysis					From the Hardware implementation				
		Power loss (W)	BT (mins)	Energy loss (Wh)	Energy efficienc y (%)	Cost (USD)	Power loss (W)	BT (mins)	Energy loss (Wh)	Energy efficienc y (%)	Cost (USD)
60	1	0.30	70	0.35	99.79	0.05	0.35	66	0.39	99.77	0.06
	13	3.90	70	4.55	97.31	0.62	4.55	66	5.01	97.00	0.71
250	1	0.33	397	2.18	98.71	0.29	0.39	365	2.37	98.59	0.33
	13	4.30	397	28.45	83.15	3.79	5.07	365	30.84	82.42	4.34

Table 3.2 Difference between experimental and simulation results derived from Table 3.1

$\Delta V(mV)$	ΔPower loss (W)	$\Delta BT(mins)$	ΔEnergy loss (Wh)	ΔEnergy efficiency (η)
60	0.05	4	0.04	0.02
	0.65	4	0.46	0.31
250	0.06	32	0.19	0.12
	0.77	32	2.39	0.73

The balancing system loses 0.3 to 0.4 W per cell and up to 5 W in the worst-case scenario (old cells) with a cell voltage of 4.2 V. Additionally, it is noted that old cells experience a considerable loss of energy. The above (Table 3.1) comparison measurement has been done for full simulation and actual system reading taking into account the unit cost of 0.13 USD/kWh in order to evaluate the cost due to energy loss in the cell balancing system (average US domestic tariff rate). Based on Table 3.1, Table 3.2 displays the differences and percentage errors between simulation and experiment. It has been noted that there are less than 1% of percentage errors between the simulation and experiment.

3. 6 Conclusion

One of the key procedures for obtaining precise battery condition estimation is battery modelling. Depending on the type of application, many internal and external factors may influence the battery SOC and the SOC calculation approaches. For the cell balancing study, the 3RC ECM is taken into consideration because it better reflects the dynamic properties of the battery than lower order models. Following chapters will implement the balancing system using the chosen 3RC battery model and its parameter values as determined by EIS test measurement.

CHAPTER 4

Active Cell Balancing and Balancing methods Comparative Study for Lithium-ion batteries in Electric Vehicles

4.1 Introduction

This chapter presents different active balancing methods relating to its application, cost, size, balancing speed, and efficiency to enhance the life cycle of the battery module. It evaluates and compares passive balancing systems with respect to commonly used active balancing systems based on inductors to choose a suitable balancing system addressing balancing speed and battery efficiency for the EV segment (E-truck, E-car, and E-bike). For efficient balance and improved accuracy, the final voltage-based balancing algorithm is used. The balancing system's most significant characteristics, which influence battery efficiency and performance, such as power loss, temperature variation, and degree of imbalance, are determined. For the optimum usage of the battery pack and to enhance the battery life, the balancing system parameters are fine-tuned using a MATLAB simulation system. Simulation results are compared with balancing parameters to find the suitable method for the EV segment. Although this research aims to optimize the passive-based approach, when the cost of the active circuitry becomes reasonable, this work can be extended to active-based algorithm as well.

4.2 Active cell balancing methods

In the active topology, the excess energy is transferred from the overcharged cell to the undercharged cells using energy storage elements [Caspar, 2018]. Fig. 4.1 represents the various cell balancing methods for the battery-powered system.

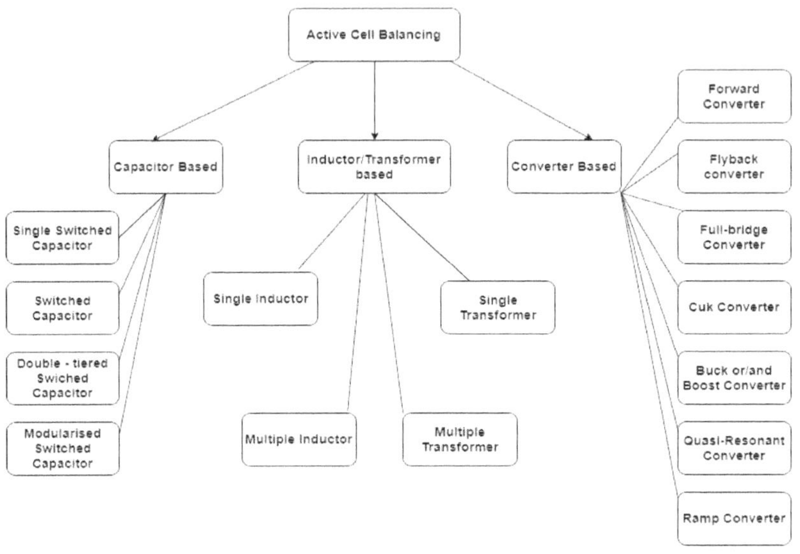

Figure 4.1 Active balancing topologies

Table 4.1 Comparison of balancing schemes from the literature [Daowd, 2011], [Kim, 2014], [Casper, 2018], [Omariba, 2019], [Andrea, 2010]

Balancing attributes	Passive balancing	Balancing using Capacitor	Balancing using Inductor	Balancing using Transformer
Preferred Mode	Charging	Charging, discharging	Charging	Charging
Balancing speed	Slow	Slow	Fast	Faster
Size	Small	Medium	Medium	Bulky
Cost	Economical	Expensive	Moderate	Very expensive
Modularization	Simple	Moderate	Moderate	Difficult to Modularize
Efficiency	> 90%	< 50%	> 95%	80% to 95%
Control complexity	Simple	Complex	Moderate	Complex
Implementation	Simple	Complex	Complex	Complex
Merits	Easy to realize Low cost	Simple control, closed-loop control is not required	Faster balancing, good efficiency, relatively cheap	Faster balancing, low switch current/ voltage stress
Demerits	Energy is dissipated as heat, slow balancing	Slow balancing, a more significant number of switches	For high switching frequency, filtering capacitors are required	Size, high cost, complex control

Table 4.1 compares the different cell balancing schemes available from the literature. A balancing scheme must enhance the power and energy stipulations, life cycle extension, and energy efficiency of the battery system.

4.2.1 Capacitor based balancing

In capacitor topologies, the capacitors move extra energy from the overcharged cells to the pack's low-charged cells. Thermal losses are generally lower when compared with the passive techniques because the capacitors transfer the cell energy. In any case, this circuit requires an intricate switching pattern and a complex switching control technique to attain balancing [Daowd, 2012].

4.2.1.1 Single switched capacitor

A single switched capacitor present in the pack is charged or discharged to balance every cell in the pack. The single switched capacitor requires an (n + 5) bi-directional switch and a capacitor to equalize 'n' cells, making the system more cost-productive and straightforward. Fig.4.2 represents the single capacitor switched cell balancer.

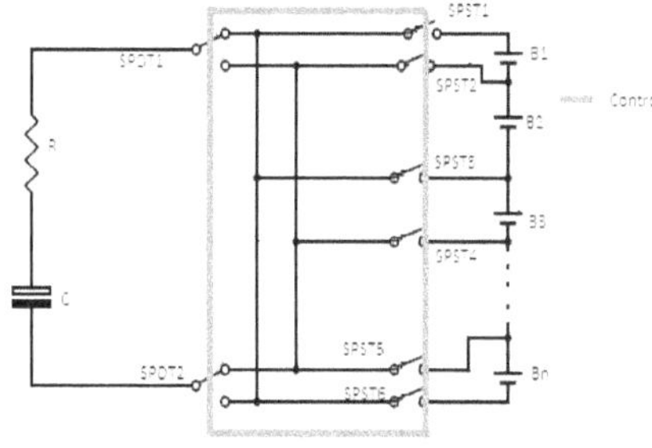

Figure 4.2 Single switched capacitor

Because of only one capacitor usage, it takes more for balancing, and this circuit needs a higher number of switches and complex control of the switching patterns

[Daowd, 2013]. This method is not suitable for battery cells, which have a minimum degree of voltage imbalance.

4.2.1.2 Switched capacitor

A single switched capacitor utilizes the single capacitor for the energy transfer. In contrast, switched capacitor uses more than one capacitor to transfer energy between adjacent cells, as shown in Fig. 4.3. It requires (2xn) bi-directional switches and capacitors of (n-1) to equalize 'n' cells [Daowd, 2012] and transfers energy between adjacent cells.

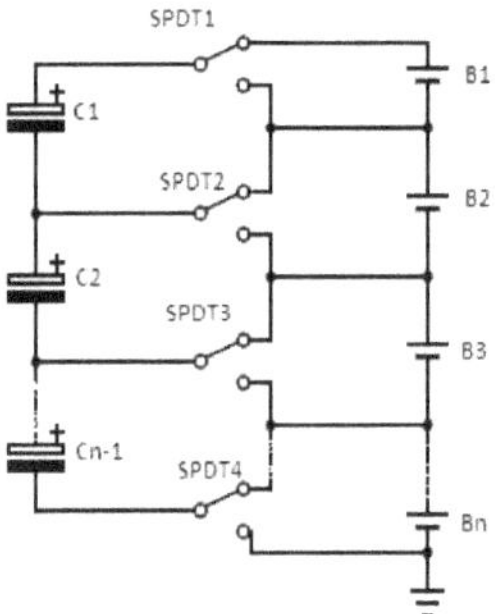

Figure 4.3 Switched capacitor

Furthermore, it does not need any complex control procedures, and it can work with both the charging and discharging process and performs well with high effectiveness. The demerits of this methodology are comparatively high balancing time and cost-inefficient compared to the passive scheme. A chain structured switched capacitor topology [Kim, 2014], series-parallel switched capacitor balancer [Ye, 2015], and star-delta switched capacitor [Shang, 2019] is introduced to reduce the balancing time for various speed requirements.

4.2.1.3 Double-tiered switched capacitor (DTSC)

DTSC topology requires (2xn) switches and 'n' capacitors for balancing 'n' cells, as shown in Fig.4.4. Having more tires provides a large number of paths between batteries, as a result, low impedance for carrying charge along a particular length across the battery pack. Balancing time is reduced to less than half, which is the

70

merit of this method over the switched capacitor topology. Besides, it operates in high efficiency while charging and discharging. The resonant topology was introduced [Ye, 2017] to overcome the slow balancing speed for a greater number of series connected cells.

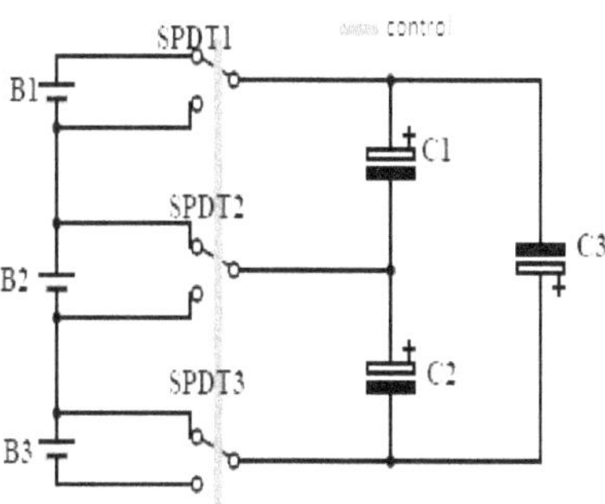

Figure 4.4 Double-tiered switched capacitor

4.2.1.4. Modularized switched capacitor

As depicted in Fig. 4.5 [Daowd, 2013], it involves battery pack modularization, the division of the battery string into groups or modules, and shuttling capacitor approaches.

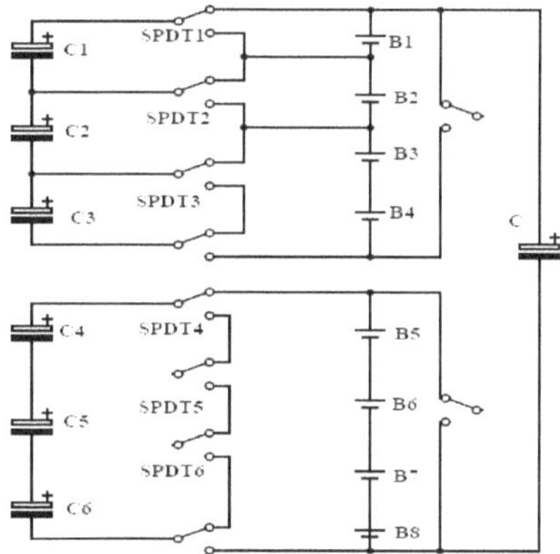

Figure 4.5 Modularized switched capacitor

Inside this topology, in each group, there is a separate balancing system dedicated to each cell, while other balancing system functions on the module level. Hence, decreasing the balancing time, increasing the capacitor count, and switch to balance 'n' cells. The demerits of this topology are a cyclic effect that, when the capacitor and switch count is increased, it leads to a rise in losses and cost of the balancing system. Resonant switched capacitor topology is used [Moghaddam, 2018] to reduce the conduction and switching losses.

4.2.2 Inductors/Transformers based balancing

Inductor/transformer cell balancer transfers energy among cells using magnetic components such as transformers and inductors. The additional charge from highly charged cells is transferred to other cells and between modules via these magnetic components. Since it uses a high balancing current, the balancing time is significantly reduced. However, this balancing circuit operates at a higher frequency, and the battery pack should have a filter component. High system costs and magnetic losses should be considered while designing this balancing scheme.

4.2.2.1 Single inductor

Energy transmission from the overcharged cell to the undercharged cell takes place through a single inductor in the buck-boost based balancing scheme. Multiple inductor cell balancing topology is used to balance the many cells simultaneously.

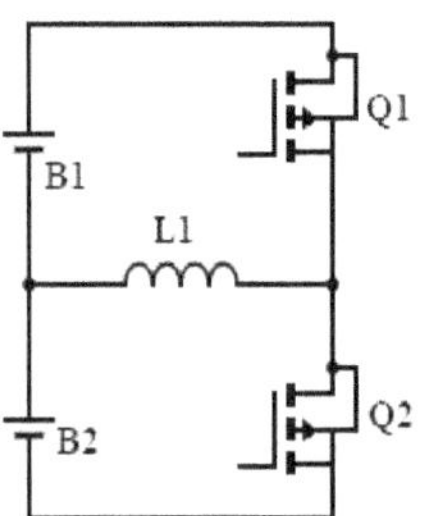

Fig. 4.6 Single inductor cell balancing

It uses a single inductor for energy transfer among the pack. The inductor is placed across the cells is charged by the excess energy in the cell while the corresponding switch is closed. This energy is then transferred to the cell with less energy as shown in Fig.4.6. Balancing time is less than capacitor-based topologies [Vardhan, 2017].

4.2.2.2 Multiple inductors

The cell balancing topology using multiple inductors, as shown in Fig.4.7, used to balance more cells simultaneously. The charge is transmitted from any cell to its neighbouring cell using this architecture. In this design, the surplus energy of the high-energy cell is transmitted through a switch to the low-energy cell by employing an inductor, which senses the voltage difference between two nearby cells. The cell count decides the balancing speed of this topology in the pack. Accurate voltage sensing is essential for fast balancing.

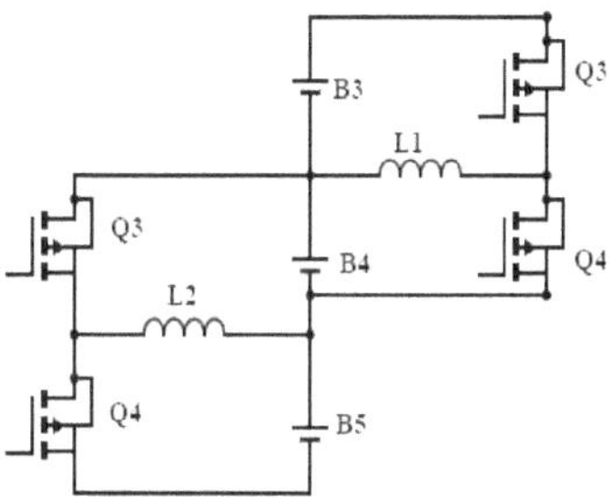

Fig. 4.7 Multiple inductor cell balancing

Parallel inductor architecture, any cell to any cell, and optimized cell to cell charge balancing are discussed in [Dong, 2015], [Zhou, 2016], [Moghaddam, 2018].

4.2.2.3 Single transformer

Single transformer based topologies are also called switched transformer topology, which can attain fast balancing with less magnetic losses [Shang, 2017], [Moghaddam, 2018]. It can transfer energy from pack to cell and cell to pack using the appropriate switch, as shown in Fig. 4.8. However, it needs complex switch control.

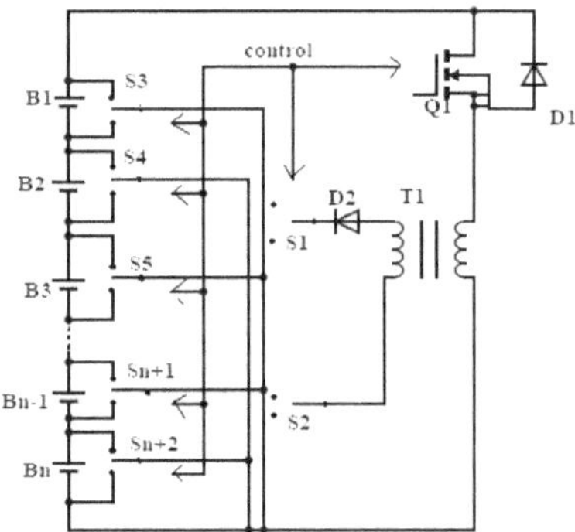

Figure 4.8 Single transformer cell balancing

4.2.2.4 Multiple transformers

Multiple transformer topologies are used to get better balancing speed, and it can balance multiple cells at the time of cell imbalance, as shown in Fig.4.9.

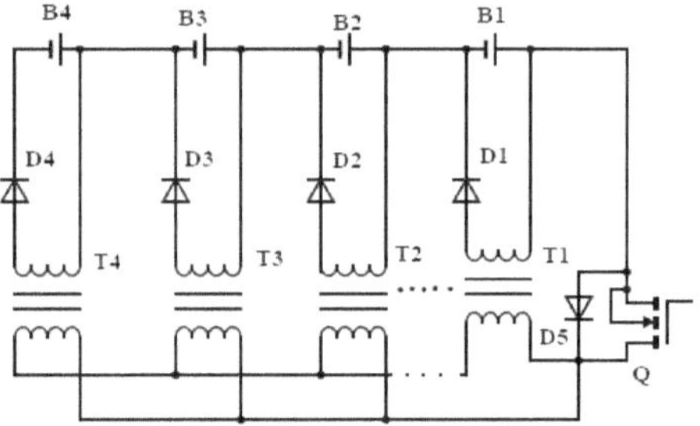

Figure 4.9 Multiple transformer cell balancing

Multiple transformer is the best topology for the modular structure with less switching stress and uses multiple transformers with individual core [Park, 2012]. Further, battery pack extension is possible by adding an extra core. However, high cost and complex circuitry are the drawbacks of this topology.

4.2.3 Converter based balancing

The dynamic cell balancing circuit, a DC-DC converter, balances the pack's cells. Topologies based on DC-DC Converters are very effective and have good control over the power flow [Amjadi, 2010]. This cell balancing circuit charge or discharge unequal charge through a transformer gives an isolated energy transmission structure. Since the magnetic losses are more in the transformer, proper circuit design is vital to increase the efficiency of this topology. It has good charge balancing proficiency compared to the dissipative method. Because each cell must be connected to additional passive components and dynamic switches, it is large and requires a complex control-switching mechanism.

4.2.3.1 Forward Converter

In the forward converters, the energy transfer occurs through transformer windings, diodes, and switches [Shang, 2017], as shown in Fig. 4.10.

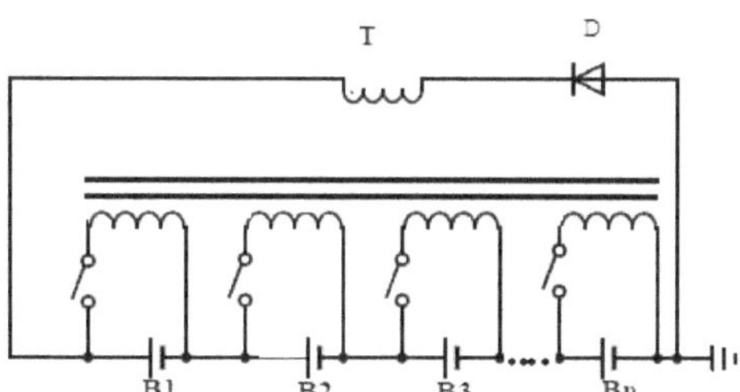

Figure 4.10 Forward converter

Forward converter experiences small voltage/current ripple and has fast balancing capabilities [Shang, 2017]. Even Though it is expensive, oversized and heavy transformer requirements are disadvantages of this topology. In addition, this topology requires a special multi-primary single secondary transformer.

4.2.3.2 Fly-back Converter

In this type of cell balancer, the transformer is used to store the estimated overvoltage on any cells, so when the switch from the auxiliary side is turned ON, energy is transferred from high voltage cells to low voltage cells through the diode [D1 to DN]. According to Fig. 4.11, this balancing system can be connected to more cells to be used for EV applications.

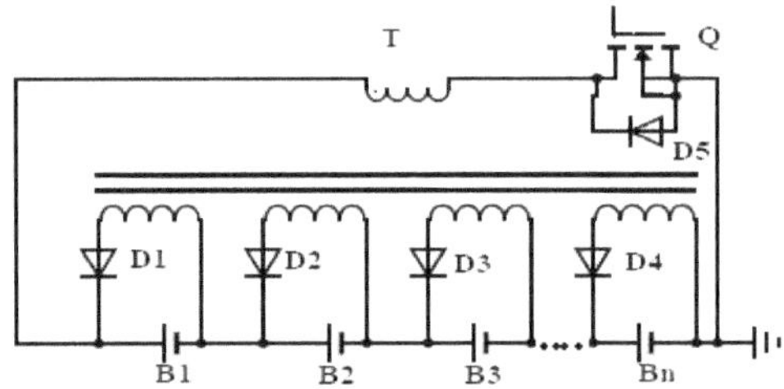

Figure 4.11 Fly-back converter

The multi winding transformer pack to cell transfer methods do, however, have a magnetic loss problem [Shang, 2017]. Every secondary side of the balancing circuit uses a unique reluctance. It costs a lot, much like the forward converter, and the turns and positions of the air gaps inside the transformer winding must be precise.

4.2.3.3 Full-bridge converter

Full-bridge converter operates buck and boost mode called fully controlled converter as shown in Fig. 4.12.

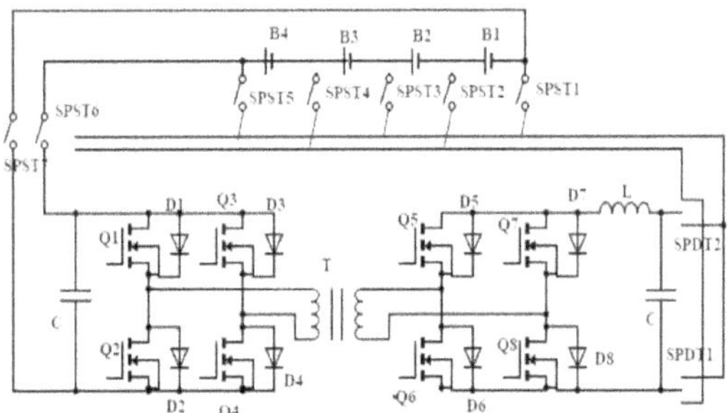

Fig. 4.12 Full bridge converter

Energy storage applications and PHEV are the systems that deploy full-bridge converters. High balancing speed and efficiency are the benefits [Chatzinikolaou, 2016]. The disadvantages of this converter are costly and complicated control schemes.

4.2.3.4 Cuk converter

The cuk converter increments or decrements the output voltage and balances only the neighbouring cells [Moghaddam, 2019], as shown in Fig.4.13.

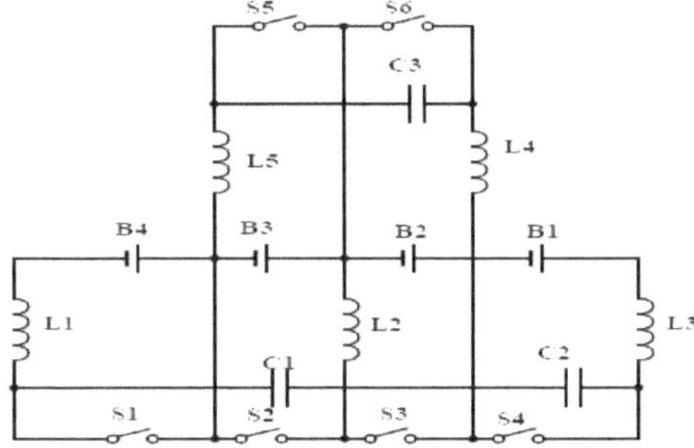

Figure 4.13 Cuk converter

The capacitors and inductors in the circuit reduces the voltage ripples. As the cuk converter exchanges the charge among two neighbouring cells, it generally requires a high balancing cost, particularly when the pack contains multiple cells [Moghaddam, 2019]. Cuk converter has a high current capacity, which allows quick balancing for EV application with extended balancing speed, but it has complex circuits to control.

4.2.3.5 Buck and/or boost converter

The buck-boost topology is preferred when the applications require high balancing currents and the cell voltage difference is minimal [Shubiao, 2017].

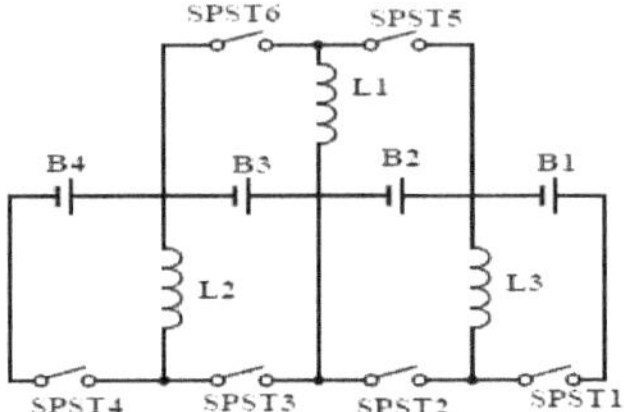

Figure 4.14 Buck or/and Boost converter

As per Fig. 4.14, the boost converter moves extra charge from a weak cell back to the battery string while the buck DC converter moves charge from the battery string to the weak cell. The weak cells are the ones that have quick charging and discharging nature due to their aging. It has the lowest energy loss and minimum balancing time compared to other converters.

4.2.3.6 Quasi-resonant converter

It is a highly efficient converter utilized for cell balancing because of less switching losses due to ZVS and ZCS operation [Shang, 2015], as shown in Fig. 4.15. Quasi-resonant circuit minimizes the switching loss, which improves the balancing efficiency. Like buck-boost topology, the Quasi-resonant converter has 96% efficiency compared to standard buck-boost, good balancing speed, small size, and generates less heat [Lee, 2015]. Disadvantages of the resonant converters are control complexity, difficulty in implementing the circuit, and high circuit cost.

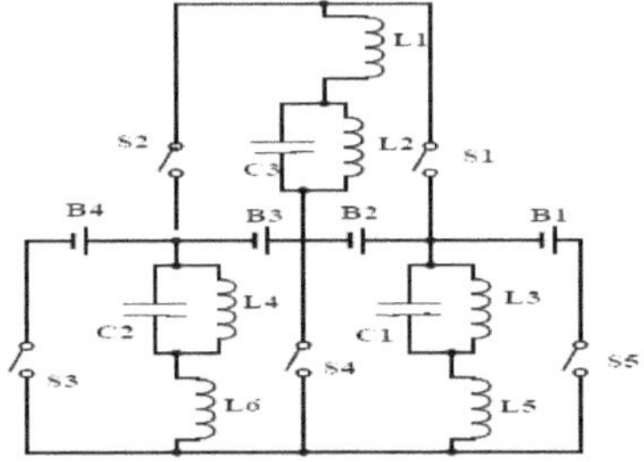

Figure 4.15 Quasi-resonant converter

4.2.3.7 Ramp Converter

Ramp Converter is similar to multiple winding transformer topology with one additional winding for the pair of cells [Gotwald, 1997] is shown in Fig.4.16.

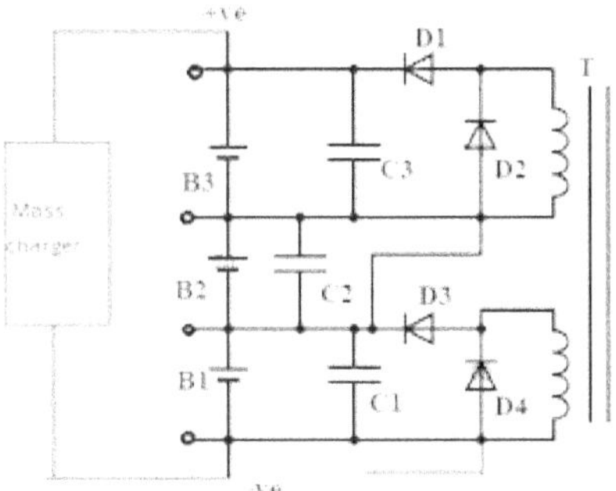

Figure 4.16 Ramp Converter

Along with action, on one-half cycles, the odd-numbered low voltage cells are charged by a more significant portion of the current, as on the next half cycle, the even- numbered low voltage cells are charged by a larger portion of the current using the ramp.

4.3 Comparison of active balancing methods

Table 4.2 contrasts the various active balancing techniques with regard to reliability, control strategy, balancing speed, and other factors. The selection of the active cell balancing topology is mostly influenced by the control strategy and design philosophies with minimal hardware components. Methods of activity and reliability, which include charge and discharge control, are used to identify control schemes. Despite the capacitor based balancing systems being used during charging and discharging modes, they are inherently 50% efficient. In addition to this, it is also not suitable for Li-ion battery chemistry, which has a minimum degree of voltage imbalance. The transformer based balancing system is less efficient than the inductor-based system and more complex. The inductor-based topology is widely used for cell balancing [Moghaddam, 2018].

Table 4.2 Comparative study of active balancing [Daowd, 2011], [Omariba, 2019], [Andrea, 2010], [Moghaddam, 2018], [Gotwald, 1997], [Lee, 2015], [Shang, 2017], [Casper, 2018]

Type	Balancing speed	Reliability	Control strategy	Cost	Size	%η	Application
Single switched capacitor	Low	Medium	Hard	High	Bulk	Low	++/+++
Multiple Capacitors	Very low	Medium	Moderate	Medium	Moderate	Low	++/+++
Single/Multiple Inductors	Medium	Medium	Medium	High	Moderate	High	++/+++
Single Transformer	Low	Low	Hard	High	Moderate	Medium	++
Multiple Transformers	Low	Low	Moderate	High	Compact	Medium	++
Forward Converter	Medium	Medium	Hard	Medium	Moderate	Medium	++
Fly back converter	Low	Low	Moderate	Medium	Moderate	Medium	++
Full bridge converter	High	Medium	Hard	Low	Compact	High	++
Cuk converter	Medium	Medium	Hard	Medium	Moderate	Medium	++/+++
Buck or/boost converter	High	High	Hard	Medium	Moderate	High	++/+++
Quasi-resonant converter	Low	Very low	Hard	Low	Moderate	Medium	++/+++
Ramp converter	Low	Very low	Hard	Low	Moderate	Low	++

+++ - High power, + - Low power, ++ - Medium power

4.4 Inductor based active cell balancing

Voltage balancing techniques are necessary for balancing system algorithms. It determines if the voltage difference between any two cells is greater than the predetermined threshold. While an active system uses an inductor to transfer extra energy to low-charged cells, a passive system uses a resistor to remove surplus energy from an overcharged cell. The balance threshold may need to be adjusted further depending on the needs of the system and the accuracy of the sampling circuit. Li-ion NMC cells have been tested, and in Chapter 2 their parameters were calculated. A battery model is drawn by the 3RC ECM approach in Chapter 3. The cell balancing is performed by using a model-based simulation in a Matlab-Simscape (R2019b). According to SOC and temperature, this model describes the OCV, SOC, and internal resistance of the battery. The thermal block is used to model the internal battery temperature, and this model successfully reflects the dynamics of the battery.

The cell balancing algorithm is set to activate with a cell threshold voltage of 3.9 V and a balancing threshold of 30 mV. The cell voltages vary with the cells' voltage deviation between 30 mV and 250 mV. Working principle, balancing architecture, modes of operation, control flow algorithm, and simulation outcome of inductor cell balancing system explained as follows.

For a 48V Li-ion battery pack with sixteen cells connected in series, the inductor-based active balancing architecture is implemented. To transmit the charge from the first cell to the last cell in a standard inductor architecture, energy must travel through each cell in the pack. As a result, the standard topology's balancing speed is inadequate. The balancing topology proposed by [Moghaddam, 2018] includes fewer switches, enhancing efficiency and minimizing power loss. The inductor topology represented in Fig. 4.17 consists of 'n' cells, (2n-2) switches, and (n-1) inductors; hence, for 16 cells, 30 switches and 15 inductors are utilized.

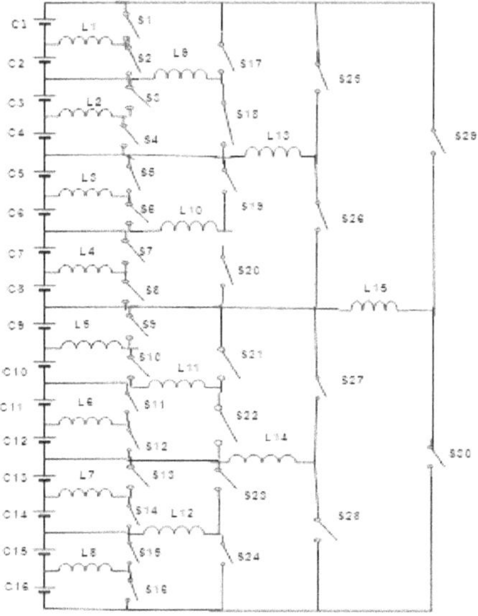

Figure 4.17 Inductor topology [Moghaddam, 2018]

Control methodology

Fig. 4.18 depicts the control logic suggested in this topology. Based on the voltage differential and balancing threshold of the cell, the process of acquiring the overcharged cells and performing cell balancing is carried out.

Stage 1

The overloaded cell is turned ON to charge the inductor if the voltage differential between the odd-numbered cells (C1, C3, C15) and their neighbouring even cells (C2, C4,.... C16) are greater than the threshold voltage (30mV). The energy stored in the inductor is transferred to the low charged cell through the body diode of the switch whenever the voltage difference falls below the threshold, turning the switch OFF. Eqn shows the voltage across the inductor L1 (4.1).For "j" series cells, the kth inductor voltage is shown in Eqn. (4.2).

$$V_{L1} = L_1 \left(\frac{di_{L1}}{dt} \right) = V_{cell1} \qquad (4.1)$$

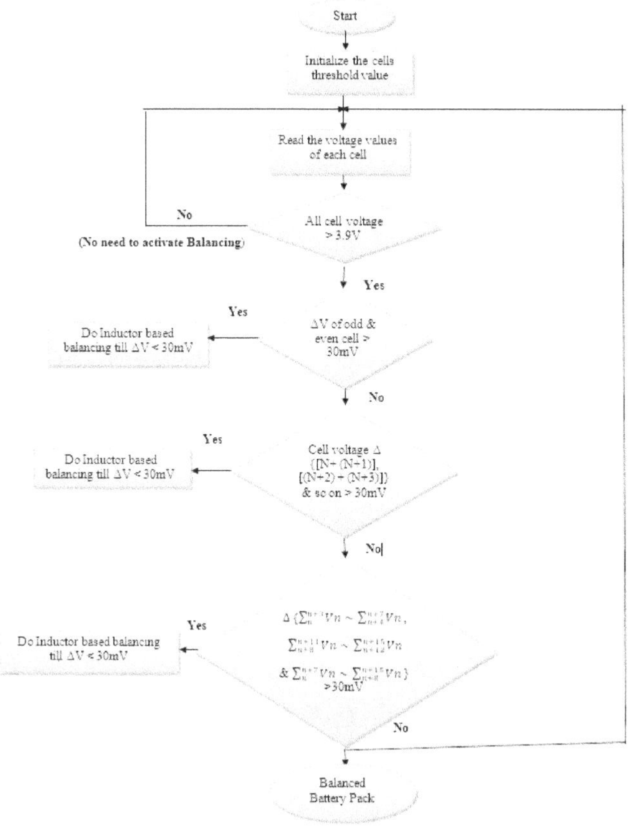

Figure 4.18 Active balancing control flowchart

$$V_{Lk} = L_k \left(\frac{di_{Lk}}{dt} \right) = V_{cellj} \tag{4.2}$$

Where, j = 3, 5......15, k = 2, 3.....8

Odd-numbered switches from S1 through S15 are turned ON, charging the inductor, if the voltage differential between odd- and even-numbered cells is larger than 30mV. If it is below threshold, the switches are turned OFF, as illustrated in Eqn. (4.3) and (4.4) show the inductor energy is transmitted to the even cells.

$$V_{L1} = L_1 \left(\frac{di_{L1}}{dt}\right) = -V_{cell2} \tag{4.3}$$

$$V_{Lk} = L_k \left(\frac{di_{Lk}}{dt}\right) = -V_{cellj} \tag{4.4}$$

Where, j = 4, 6......16, k = 2, 3.....8

Stage 2

The sum of two cell voltages $(V_1 + V_2)/2$ is compared with the next two cell voltages $(V_3 + V_4)/2$ if it is more than 30mV, the overcharged cell is turned ON. The same strategy is applied to other cells. The voltage across the inductor L9 and Lk is shown by the Eqn. (4.5) and (4.6)

$$V_{L9} = V_{cell1} + V_{cell2} = L_9 \left(\frac{di_{L9}}{dt}\right) \tag{4.5}$$

$$V_{Lk} = L_k \left(\frac{di_{Lk}}{dt}\right) = V_{cellj} + V_{celli} \tag{4.6}$$

Where, i = 6, 10, 14, k = 10, 11, 12, j = 5, 9, 13

The switches are turned off if the cell voltages are below threshold, the inductor energy is transferred to the low charged cells is shown in Eqn. (4.7) and (4.8).

$$V_{L9} = -V_{cell3} - V_{cell4} = L_9 \left(\frac{di_{L9}}{dt}\right) \tag{4.7}$$

$$V_{Lk} = L_k \left(\frac{di_{Lk}}{dt}\right) = -V_{cellj} - V_{celli} \tag{4.8}$$

Where, j = 7, 11, 15, k = 10, 11, 12, and i = 8, 12, 16

Stage 3

The difference in cell voltages of (V1+V2+V3+V4)/4 is compared with cell voltages of (V5+V6+V7+V8)/4, and the cell voltages of (V9+V10+V11+V12)/4 are compared with cell voltages of (V13+V14+V15+V16)/4. If the voltage deviation is more than 30mV, the corresponding switches are turned ON. The voltage across the inductor L13 and L14 is shown in Eqn. (4.9) and (4.10)

$$V_{L13} = L_{13}\left(\frac{di_{L13}}{dt}\right) = V_{cell1} + V_{cell2} + V_{cell3} + V_{cell4} \tag{4.9}$$

$$V_{L14} = L_{14}\left(\frac{di_{L14}}{dt}\right) = V_{cell9} + V_{cell10} + V_{cell11} + V_{cell12} \tag{4.10}$$

The energy stored in the inductor is moved to the low-charged cells in the next mode is shown in Eqn. (4.11) and (4.12).

$$V_{L13} = L_{13}\left(\frac{di_{L13}}{dt}\right) = -V_{cell5} - V_{cell6} - V_{cell7} - V_{cell8} \tag{4.11}$$

$$V_{L14} = L_{14}\left(\frac{di_{L14}}{dt}\right) = -V_{cell13} - V_{cell14} - V_{cell15} - V_{cell16} \tag{4.12}$$

Stage 4

In the fourth stage, the voltages of the first eight cells and the last eight cells are compared. Depending on the voltage difference, the matching switches are switched ON. Eqn. (4.13) shows the voltage across the inductor L15.

$$V_{L15} = V_{cell1} + V_{cell2} + V_{cell3} + V_{cell4} + V_{cell5} + V_{cell6} + V_{cell7} + V_{cell8} =$$

$$L_{15}\left(\frac{di_{L15}}{dt}\right) \tag{4.13}$$

The energy stored in the inductor L15 is transferred to the bottom eight cells in the following mode is shown in Eqn. (4.14),

$$V_{L15} = L_{15}\left(\frac{di_{L15}}{dt}\right) = -V_{cell9} - V_{cell10} - V_{cell11} - V_{cell12} - V_{cell13} - V_{cell14} -$$

$$V_{cell15} - V_{cell16} \tag{4.14}$$

This selected inductor topology can balance top, centre, and bottom cells with high balancing speed.

4.4.1 Cell balancing design

Energy efficiency is a key feature when designing an energy storage system. Optimal dimensioning of the balancing system achieves higher energy efficiency when balancing, saving valuable energy instead of wasting it in heat. Both passive and active balancing system implementations require more resistors, inductors, and MOSFET switches in the balancing circuit. Each minor configuration for a certain balancing current has the potential to alter the volume of the balancing architecture and energy dissipation.

Table 4.3 Properties of MOSFET & inductor

MOSFET	Symbol	Inductor	Symbol
Max. Current	I_{max}	Max. Current	I_{max}
Switch resistance	R_{DS}	Inductance	L
O/P Capacitance	C_{oss}	DC Resistance	R_{ind}
Delay time	t_{on}, t_{off}	Volume	v_{ind}
Volume	v_m		

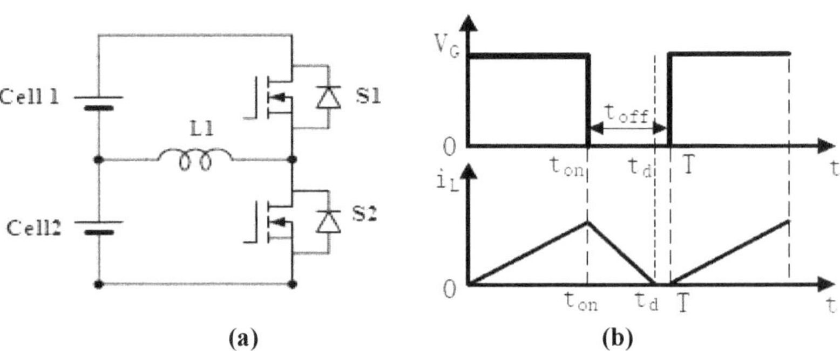

Figure 4.19 (a) Balancing operation, (b) DCM mode

The properties of MOSFET and inductors, which influence the balancing parameters of the balancing design, are listed in Table 4.3 [Narayanaswamy, 2014]. In the second step (with the switch open), the inductor's accumulated energy is transferred to the low voltage cell (switch opens and diode works). The duty cycle

D should be < 0.5 cycles; hence, the balancing inductor works in discontinuous current mode (DCM). The DCM mode of DC-DC converter operation is represented in Fig.4.19a and b [Wang, 2017]. When switch S1 is ON, the cell1 voltage charges the inductor L1 is shown in Eqn. (4.15).

$$V_{Lon} = i_{Lon} R_{0n} + L1 \frac{di_{Lon}}{dt} + V_m , 0 < t \leq t_{0n} \tag{4.15}$$

Where,

- V_{Lon} - voltage of the inductor L1
- i_{Lon} - current through the inductor
- R_{0n} - resistance in the current path (cell internal resistance, wire resistance, inductor equivalent series resistance, etc.)
- V_m – voltage of the MOSFET

When switch S1 is turned OFF, the inductor's stored energy is transferred to cell 2, and V$_L$off, which stands for the voltage across the inductor, is represented in Eqn (4.16).

$$V_{Loff} = i_{Loff} R_{0ff} + L1 \frac{di_{Loff}}{dt} + V_d, t_{0n} < t \leq t_d \tag{4.16}$$

Where,

- V_d - forward voltage drop of the MOSFET diode,
- t_d – time when the inductor current drops to zero.

Since R_0 is <<, neglecting R_0. The ON and OFF time of the switch is calculated as follows,

$$t_{on} = \frac{L1. \Delta i_{Lon}}{(V_{Lon} - V_m)} \tag{4.17}$$

$$t_{off} = \frac{L1. \Delta i_{Loff}}{(V_{Loff} - V_D)} \tag{4.18}$$

The peak value of the inductor current during switch ON and OFF is calculated from Eqn. (4.17) and (4.18).

$$\Delta i_{Lon} = \frac{(V_{Lon} - V_m)t_{on}}{L1} \tag{4.19}$$

$$\Delta i_{Loff} = \frac{(V_{Loff} - V_D)t_{off}}{L1} \tag{4.20}$$

By ignoring the drop across the switch from Eqn. (4.19) and diode drop from Eqn. (4.20), the value of peak inductor current during switch ON and OFF as shown in Eqn. (4.21) and (4.22)

$$\Delta i_{Lon} = \frac{(V_{Lon}t_{on})}{L1} = \frac{(V_{Lon})D.T}{L1} \tag{4.21}$$

$$\Delta i_{Loff} = \frac{(V_{Loff}t_{off})}{L1} = \frac{(V_{Loff})(1-D).T}{L1} \tag{4.22}$$

The inductor current is calculated by using Eqn. (4.23),

$$I_L = \frac{(V_{Lon})D}{R(1-D)^2} \tag{4.23}$$

The value of the minimum and the maximum current is calculated by using the Eqn. (4.24) and (4.25).

$$I_{max} = I_L + \frac{\Delta i_{Lon}}{2} \tag{4.24}$$

$$I_{min} = I_L - \frac{\Delta i_{Lon}}{2} \tag{4.25}$$

Substituting Eqn. (4.21) and (4.23) in Eqn. (4.25) and equating it to zero, the minimum value of the inductor is calculated,

$$L_{on} = \frac{R(1-D)^2}{2f} \tag{4.26}$$

$$i_{Bavr} = \frac{(V_{Lon}).D^2}{2L1f} \tag{4.27}$$

The lowest value of the inductor is determined by Eqn. (4.26), and the average value of the balancing current is determined by Eqn. (4.27). The switching frequency and inductance have an inverse relationship with the average value of the balancing current. Since the balancing time has an inverse relationship with the balancing current, the balancing time is dependent on f and L. The inductor size can be minimized further by increasing the switching frequency.

4.5 Simulation outcomes

The chosen model is simulated using Matlab/Simscape under charging settings to evaluate how well the framework works. A look-up table contains the parameter values for the model. Cells are charged in CC-CV mode, and the voltage variation across the cells is calculated by altering the battery pack's initial state of charge (SOC). The cell has a 3.3Ah capacity, and a 10µH inductor value has been chosen. Calculations are made for each cell's OCV, terminal voltage, temperature, and SOC.

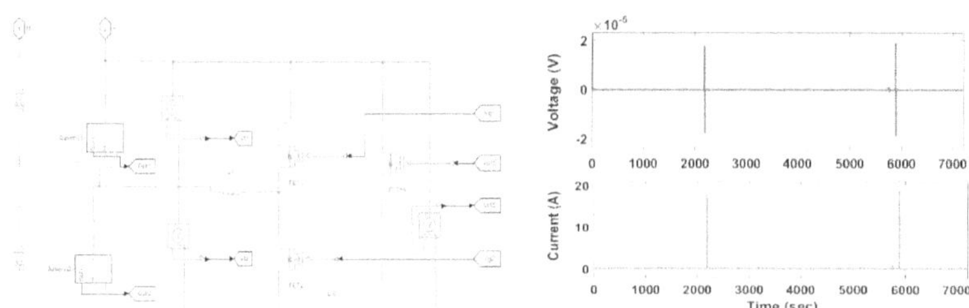

Fig 4.20 Cell1 and 2 architecture in Matlab   **Fig 4.21 Voltage and current of inductor1**

Initially, the voltage deviation between cell1 and cell2 is kept above 30mV. When the cell voltages hit the threshold charge voltage (3.9V) while the cells are being charged at a rate of 0.5C, the cell balancing is initiated. The Matlab/Simscape structure of cell1 and cell2 is shown in Fig 4.20.

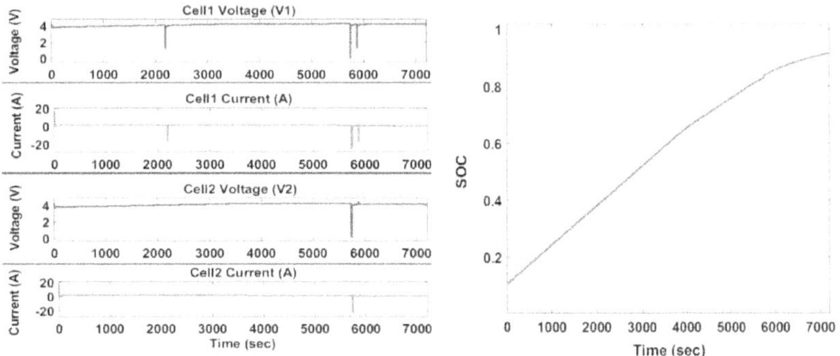

Figure 4.22 Voltage and current of cell1 and 2   **Figure 4.23 Battery SOC**

Fig. 4.21 shows the current and voltage charts of the inductor1 during balancing. Stage 1 begins with switch S1 being switched ON as soon as cell 1 voltage exceeds cell 2 voltage. Switch S1 then charges the inductor 1 and is kept ON until the voltage difference between the cells is less than 30 mV. The energy from the inductor 1 is transmitted to cell 2 at 5,800 seconds via the switch S2 body diode. The cell voltage fluctuations in Fig. 4.22 are negligible due to the cells' low balancing threshold voltage. According to Fig. 4.23, the charging process takes around two hours to complete, and the cell balancing takes about 50 minutes.

The charging occurs at an ambient temperature of 20°C. The cell temperatures increase up to 24.75°C during charging, as shown in Fig. 4.24. In stage2, cells 1-2 voltages and cells 3-4 voltages are compared, and so on. In stage3, cells 1-4 and 5-8 voltages are compared. During stage4, cells 1-8 voltages are compared with cells 9-16 voltages.

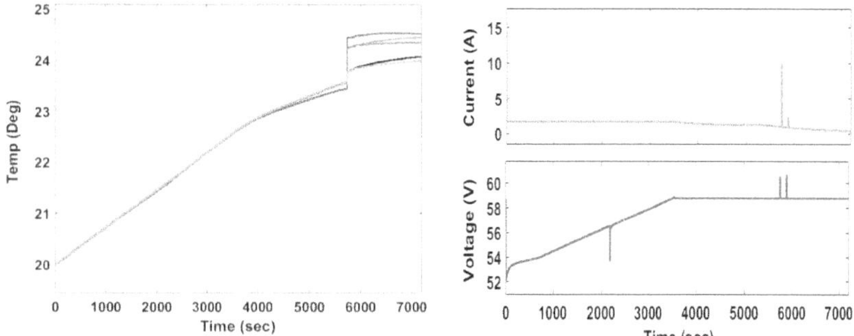

Figure 4.24 Battery temperature   Figure 4.25 Charging voltage and current

Although the voltage varies while the batteries are charging, the current stays constant. When the cell reaches its maximum voltage, it is switched to the constant voltage mode, as illustrated in Fig. 4.25. In this mode, the voltage is maintained constant while the current reduces further to the lowest base value permitted by the power rating.

The passive and active simulation results show that the passive balancer takes more time due to higher voltage deviation among the cells, as shown in Fig.4.26 and 4.27. Active balancing overcomes the limitations of passive balancing, as given in Fig.4.28 and 4.29. The heat dissipation across the balancing system could be the main reason for the temperature rises near the battery pack. It reduces battery life by stimulating the aging mechanisms.

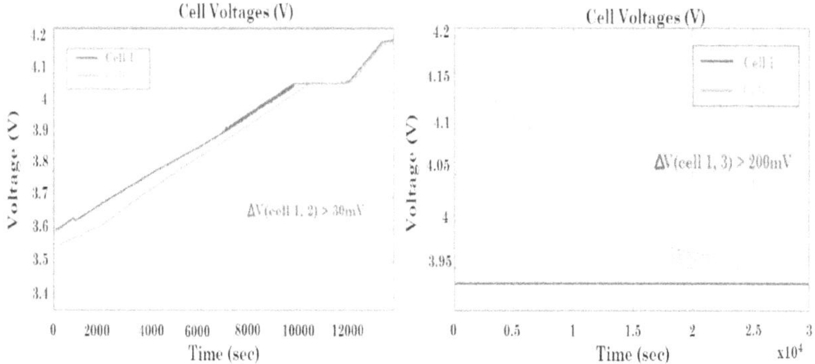

Figure 4.26 Passive balancing (charging)   Figure 4.27 Passive balancing (idle)

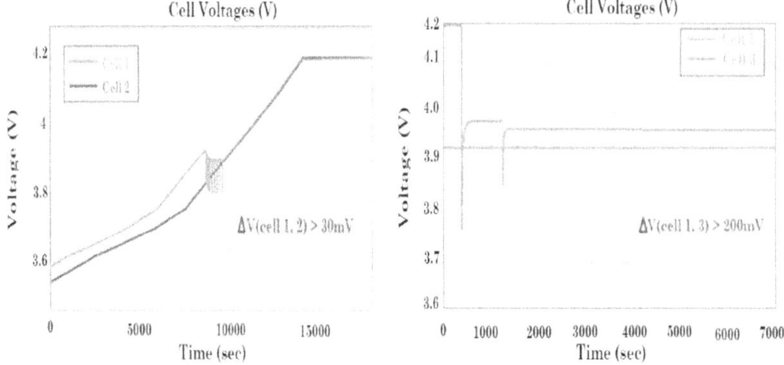

Figure 4.28 Active balancing (charging)   **Figure 4.29 Active balancing (idle)**

The balancing system's temperature characteristics are studied by evaluating the maximum temperature (T_{max}) and temperature distribution (ΔT) among the cells. The active system reaches a lower maximum temperature (24 °C) than the passive system (27 °C), and ΔT does not exceed 4°C at the time of charging, as shown in Fig.4.30 and 4.31.

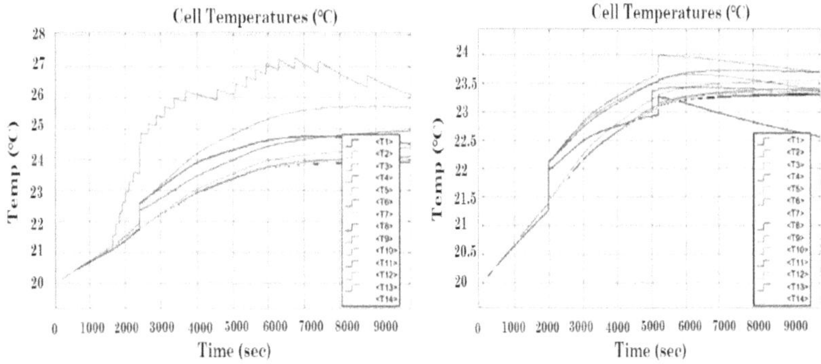

Figure 4.30 Temperature rise (passive)   **Figure 4.31 Temperature rise (active)**

92

Table 4.4 Passive system energy efficiency and cost calculation

ΔV (mV)	Cells (#)	P_{loss} (W)	BT (min)	E_{loss} (Wh)	η (%)	Cost ($)
30	1	0.35	48	0.28	99.83	0.04
	13	4.55	48	3.64	97.84	0.49
200	1	0.35	215	1.25	99.26	0.17
	13	4.55	215	16.30	90.34	2.17

Table 4.5 Active system energy efficiency and cost calculation

ΔV (mV)	Cells (#)	P_{loss} (W)	BT (min)	E_{loss} (Wh)	η (%)	Cost ($)
30	1	0.07	15	0.02	99.9	0.003
	13	1.05	20	0.35	99.7	0.047
200	1	0.4	35	0.23	99.86	0.031
	13	1.16	42	0.81	99.51	0.108

4.5.1 Calculation of key parameters from simulation

The simulation outcomes deliver many vital parameters, including power loss, balancing time, energy loss, energy efficiency, and total cost of the balancing system. Tables 4.4 and 4.5 show the passive and active simulation results of a typical 48 V E-bike system. Using passive balancing shows that a significant energy loss occurs only when the degree of imbalance is high. Nevertheless, the energy loss observed is low in the active topology, irrespective of the quantum of imbalance and the number of imbalanced cells in the pack.

4.6 Discussion

4.6.1 E-vehicle use cases

In low power E-bikes, since the voltage requirement is not very high, the number of cells that go out of balance is considerably less. In a fourteen-cell series connected battery pack, the passive balancer took 48 minutes to bring down the cell voltage below the balancing threshold (30 mV), as shown in Fig.4.26.

.

Table 4.6 Comparison of cell balancing system characteristics

Parameters	E-Bike		E-Car		E- Truck	
	PB	AB	PB	AB	PB	AB
Cells in series	14	14	96	96	250	250
ΔV (mV)	30	30	30	30	30	30
P_{loss} (W)	4.55	1.05	38	9.5	99.6	24.9
BT(Min)	48	20	48	20	48	20
E_{loss} (Wh)	3.64	0.35	30.4	3.2	80	8.3
η (%)	97.8	99.7	97	99	96.7	98.5
E_{loss} cost ($)	0.49	0.05	4.05	0.4	10.7	1.11

PB – Passive Balancing, AB – Active Balancing

The power dissipation of passive balancing is minimum (0.35 W), as shown in Table 4.4. For these scenarios, the selection of active balancing is not feasible due to circuit complexity and high cost. Hence, passive balancing is applicable for low power E-bikes. The passive balancer took 215 min to balance with a power loss of 4.55W for the maximum voltage deviation (aged cells) and for more imbalanced cells (cycled many times). Since the loss is significant (shown in Table 4.4), the active balancer could be a better option. The active balancer took 15 to 42 min for the difference of 30 mV to 200 mV, as shown in Table 4.5. Table 4.6 shows a comparative study of the cell balancing system characteristics of EV variants. It is observed that cost saving from the energy loss is not significant with passive balancing even though power loss is reduced considerably with active balancing. Due to the high-energy loss, the passive balancing system also produces hot spots within the pack, affecting the battery RUL and SOH.

4.6.2. Balancing under charging, discharging, and idle state

When a cell's voltage reaches the maximum voltage threshold during charging, passive cell balancing is used [Koseoglou, 2020]. As some of the battery energy is dissipated, it is not suitable other than charging time. The balancing process is finished (48 min) within the charging period (5 to 6 hr) is shown in Table 4.4. For a high degree of voltage imbalance (>200 mV), it took more time to balance (215 min). Hence, based on the imbalance, the balancing algorithm must decide when to start the balancing at either low SOC or high SOC region to balance the cells within the charging cycle. The balancing current is restricted because the maximum heat dissipation across the resistor makes the balancing process slow and creates an issue for fast-DC charging applications, whereas active balancing redistributes the extra charge and makes sure that all the cells are balanced even when the EVs are at the beginning or end of charging, parked, or driving. The active balancer took almost equal time regardless of the voltage difference, as shown in Table 4.5. Having suitable balancing current marks, it balances the cell voltages quickly, thereby giving a higher charging efficiency. Table 4.7 illustrates C-rate, time while discharging, and charging batteries of 1Ah. Numerous charging strategies have been suggested for fast charging Li-ion batteries, yet the safely recognised C-rate of EV batteries is currently limited to 1-1.5 C [Tomaszewska, 2019].

Table 4.7 C-rate vs. charging time

C-rate(C)	Time (hr)
5	0.2
2	0.5
1	1
0.5	2
0.2	5
0.1	10
0.05	20

Source: Batteryuniversity.com

4.6.3 Energy concerns

Due to power loss over the passive system balancing resistor, the active system's charging efficiency is higher than the passive systems, and it is significantly lower for high-power EVs (shown in Table 4.4). As shown in Eqn. (3.25 and 3.26), the energy loss in the passive system is estimated using cell voltage, balancing resistor, and the balancing time value. In contrast, the active system generates heat due to balancing converter power losses. Sometimes the power wastage of the active system leads to more significant loss than the passive balancing; even in standby mode [Andrea, 2010]. Tables 4.4 and 4.5 show an increase in energy efficiency by 1.86% for the maximum number of imbalanced cells and 9.17% for the cells with higher voltage differences. So, active balancing is suggested for the application, which needs a medium or large battery pack that should be highly energy efficient. The power handled during the balancing is usually minimal (0.1 to 10 W/cell), and if it is noticeable (>10W/cell), then there is a possibility that both the balancing system generates same amount of heat due to active system standby power loss [Andrea, 2010].

4.6.4 Temperature behavior

In order to improve the power capacity, Ni content in NMC cells is improved in different variants (NMC111 to NMC811). However, it increases the battery temperature and introduces safety issues due to the high reactivity of the cathode material with the electrolyte [Ma, 2016]. The NMC cathode shows a poor temperature tolerance at high temperatures during the thermal run-away test [Lei, 2017]. Packing a set of similar performance cells while making modules have shown a balanced temperature distribution and lowered temperature rise [Li, 2019]. The temperature performance of the battery pack is highly dependent on the balancing system, which controls parameters such as temperature deviation (ΔT) among the cells and the maximum temperature of the pack (T_{max}) as per the experiment conducted (shown in Fig.4.30 and 4.31). The power loss across the cell cause temperature rise in the pack. The balancing system balances the cell voltages so that the ΔT of the cells is minimized and T_{max} is low in the case of active (shown in Fig.4.31).

4.6.5 Cost analysis of the balancing system

Passive system architecture requires a power MOSFET switch S_n combined with a balancing resistor R_B to dissipate the excess energy of the cells. Active systems that have complex architecture need a more significant number of inductors and active switches. The overall cost of passive and active systems could be reduced by using specific BMS ICs. Table 4.8 shows the balancing system cost.

Table 4.8 Balancing system elements and cost (14 cells, 48V battery system)

PB				AB			
Part	#	$ / pcs	$	Part	#	$ / pcs	$
BMS IC	1	13.3	13.33	BMS IC	1	16	16.00
Resistor	14	0.14	1.96	Inductor	15	1.33	19.95

PB – Passive Balancing, AB – Active Balancing

4.6 Conclusion

There is discussion of various active balancing topologies. The inductor topology is chosen for examination based on the discussion. When compared to the conventional inductor-based topology, the chosen topology may balance the middle, top, and bottom cells concurrently. Based on the difference in cell voltage, the control techniques are applied to the switches. The benefits of this topology include quick balancing and straightforward control. The balancing algorithm contributes to reliable performance and a longer life cycle of EV battery packs. Passive balancing system finds application in low power systems (E-bike, E-rickshaw, etc.) with cells, which are closely matched and charged through the conventional charging method. Battery packs for high power applications (E-car, E-truck, etc.) contain a greater number of series connected cells, which leads to significant imbalance and hence needs a high current active balancing system. During balancing, the temperature deviation (ΔT) is less than 1°C in the active system and around 3°C in the passive system. By using active balancing, the efficiency of the battery system is improved by 2% for cells having low ΔV and about 9% for cells having high ΔV. Passive system is simple to implement and has less cost. Nevertheless, it involves high investment in the thermal management system. However, the active system is expensive and more complex; substituting the active system with the passive for high power battery applications (e.g., E-truck) reduces thermal issues while increasing energy efficiency.

CHAPTER 5

Optimized Passive Balancing Approach for Fast Charging in Electric Vehicles using Machine Learning

5.1 Introduction

This chapter describes how the addition of switched variable resistor topology improves the balancing speed of traditional passive balancing systems. A control flow technique is used to describe the workings of fixed and variable resistor topology. By taking into account the ideal balancing resistor selection based on the charging C-rate, balancing time, temperature rise, and degree of cell imbalance, the ML based cell balancing algorithm is developed to improve the power loss management and balancing time of passive cell balancing. To reduce power loss and improve thermal characterization, variable resistors are utilised. The effectiveness of the system is evaluated using BPNN, LSTM, and RBNN. Mean square error (MSE), mean absolute error (MAE), and root mean square error are used to verify the proposed system's performance (RMSE). The implemented control algorithm reduces the system's cost and circuit complexity. In Matlab, the proposed control techniques are validated.

5.2 Proposed passive system implementation

For the Li-ion cells, the suggested passive system operation mode has been divided into conventional (slow charging) and DC fast charging modes. According to an analysis of the charging protocols for three distinct Li-ion cell chemistries in [Keil, 2016], the CC-CV is appropriate for quickly charging high-power Li-ion cells. The additional charge from the overcharged cell is removed using a constant value of resistors in the conventional passive balancing method, as depicted in Fig.5.1. These systems use resistors in the 30 to 40 ohm range to accommodate the 100 mA

99

balancing currents for 4.2V cell voltage. To balance the cells, though, more is required. The system needs to balance more quickly in conditions of fast charging and maximum V variation between cells in order to shorten the balancing time by increasing the balancing current. To do this, shunt employing selection logic employs an array of resistors rather than fixed resistors. According to the necessary balancing current, as shown in Fig. 5.2, the suggested passive system selects a variable resistor from the parallel arrangement of resistors.

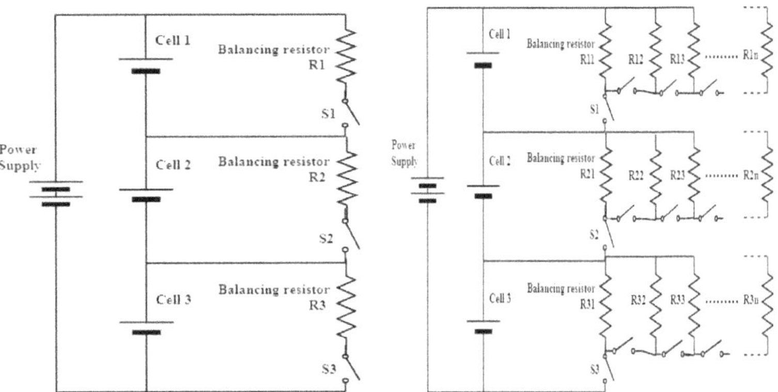

Figure 5.1 Conventional passive balancing circuit   **Figure 5.2 Proposed passive balancing circuit**

By connecting additional resistors in parallel with the existing resistor, the total value of balancing resistors is reduced. This improves the balancing current and reduces the balancing time. To lessen the Li-ion cells' inconsistency, the balance parameters must be optimised.

Battery charging time, charging frequency, charger utilisation (using one charger for multiple applications), manufacturing cost, mechanical aspects (high reliability, low weight, and low vibration), and balancing current are all factors that need to be taken into account when choosing the best passive balancing design [Vereb, 2018]. The following formula is used to determine the cell's balancing current [Andrea, 2010].

$$\text{Balancing current } (I_B) = \frac{\text{Delta Charge (Ah)}}{\text{Balancing time (h)}} \qquad (5.1)$$

The anticipated amount of balancing current is derived from Eqn. (5.1). However, because it is connected to power loss, caution should be used while choosing the balancing current. The balancing current and battery voltage are used to calculate the value of the balancing resistor. The balancing resistor selection is crucial in addressing the slow balancing and heat dissipation issues of traditional passive algorithms, according to equations 3.17, 3.18, and 5.1.

5.2.1 Balancing resistor optimization

The suggested balancing system algorithm takes into account three crucial balancing factors (power loss, temperature rise, and balancing speed) because they are connected to the value of the balancing resistor in order to address the issues with the current passive balancing systems. Think about the important run-time variables such charging frequency, battery chemistry, thermal effect, charging time, imbalance risk, charging temperature, and charger utilisation. The following are the key characteristics of the suggested passive balancing approach:

- Implementation of the best passive balancing mechanism while taking into account the varied operating and environmental conditions and battery chemistry

- The ML algorithm chooses the right resistor so that the system's balancing speed is increased without significantly raising the system's power loss or temperature.

- For validation and verification, the performance indices of the proposed ML models are evaluated based on the balancing resistor power loss, balancing time, MSE, RMSE, and MAE

The best resistor choice is important and heavily depends on the following factors.

• Charging time: The modest balancing current is enough to balance the cells when the battery is charged slowly. To balance the cells during the charging period, a large balancing current is necessary if it is charged at a high C-rate.

• Charging temperature: Because battery pack temperature rise is greater in hotter climates than it is in colder climates, the ambient temperature of the battery must also be taken into account when building the balancing resistor.

• Cost for thermal management: The investment for a thermal management system is considerable since more energy is lost in the balancing circuit due to the demands of quick charging. To maintain the thermal safety of the system, it is crucial to choose the balancing resistor using the proper control algorithms.

• Battery chemistry: Depending on the cell chemistry, the balancing current and imbalance rate change dramatically.

• Probability of imbalance: This value represents the battery pack's age and the degree of imbalance. When the cell-manufacturing batch is different, the imbalance is more obvious. However, if the production lot size is big, this can be managed. As a cell gets older, the likelihood of imbalance grows.

In a traditional passive balancing system, the maximum balancing current is fixed and cannot be changed as needed. To solve this issue, a passive balancing method based on variable resistors is presented, taking into account the balancing system's performance metrics. To deal with these issues, many optimization strategies are thoroughly examined. The battery temperature is tracked in order to prevent thermal runaway issues while the balancing time of the passive system is checked under various charge C-rates and minimum to maximum voltage imbalances (1V) of the cells. The ideal resistor choice is influenced by the battery temperature, voltage imbalance, and charging duration. Three levels of voltage imbalances are distinguished: low, medium, and high. Three cells are taken into account for the simulation study, together with two parallel configurations of resistors with various resistances, as illustrated in Fig. 5.3. The selection of the resistor value range is dependent on the battery chemistry, the BMS balancing board components' tolerance for the balancing current, and the maximum temperature limit.

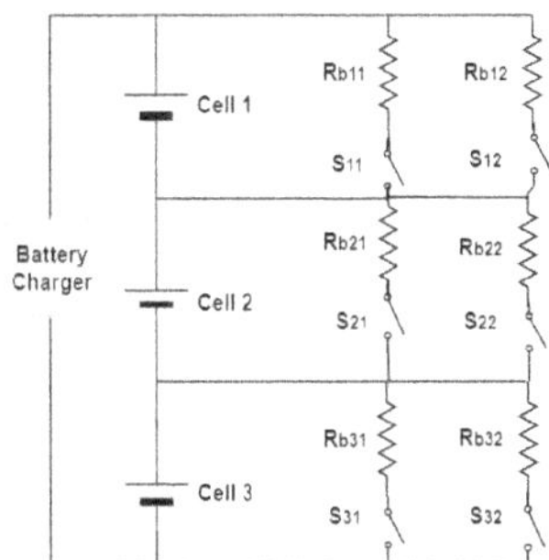

Figure 5.3 Proposed approach – dedicated pair of resistors per cell

For different balancing resistors, the balancing current consumption, power loss, and temperature increase are estimated. Based on balancing current, time, and power loss equations, as illustrated in Fig. 5.4, the range of resistor values taken into consideration for this analysis is from $1\,\Omega$ to $40\,\Omega$.

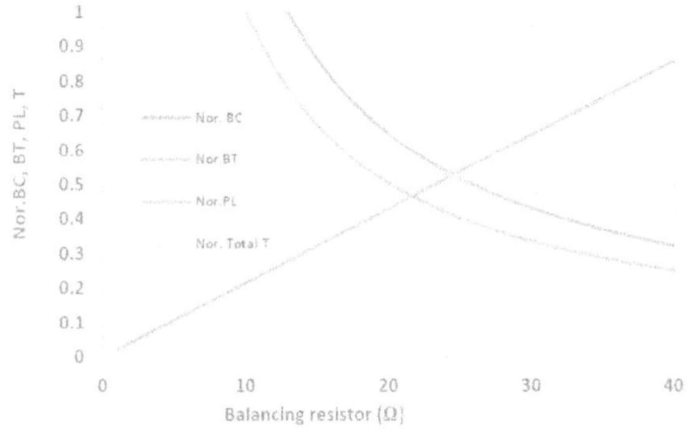

Figure 5.4 Resistor selection with respect to balancing time (BT), balancing current (BC), battery temperature (T), and power loss (PL)

In Fig. 5.5, the battery temperature is calculated by taking into account various ambient temperature conditions (0°C, 15°C, 30°C, 45°C, 60°C, 10°C, and 20°C). The calculation shows that the balancing current is within the allowed range of 300mA. The temperature and power loss are both within the permissible range for the resistor range of 15Ω to 40Ω. For Rb11, Rb21, and Rb31, the chosen resistor value is 33Ω, and for Rb12, Rb22, and Rb32, it is 25Ω.

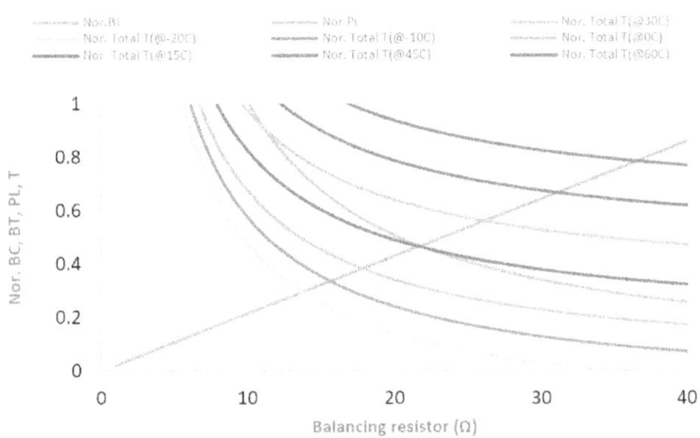

Figure 5.5 Resistor selection with respect to BT, BC, T, PL, and Ambient temperatures

Fig. 5.6 provides a description of the algorithm's design. The following circumstances are created for simulation by balancing changing voltage and current rate. The initial voltage differences are between 30 and 50 mV, 50 and 100 mV, and above 100 mV, respectively, with charging currents of 0.5 and 1 C (1.675 and 3.35 C). As long as the battery temperature is within safe ranges, both resistors are turned ON for a higher C-rate and cell imbalance (ΔV>100mV). The first resistor of the cell turns on when the charging current is less than or equal to 0.5C and the voltage differential is between 30 and 50 mV. The second resistor of the cell is turned ON if the voltage differential is between 50 and 100 mV.

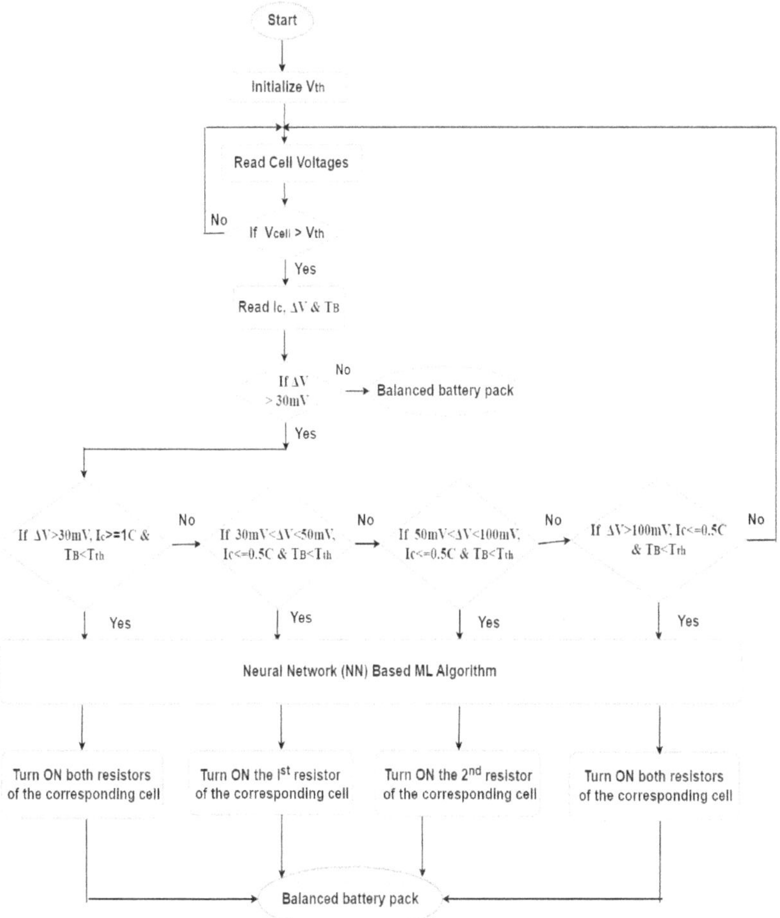

Figure 5.6 Proposed balancing control flowchart

5.3 Machine learning methodologies

In the suggested method, a machine learning (ML) algorithm is added to the improved passive balancing mechanism in order to pick the balancing resistors as best as possible given the environmental conditions and user experience requirements. The numerous ML methods utilised in the BMS application [Man-Fai, 2020] features are described in Fig. 5.7, which is a thorough classification of machine learning, but it does not answer the needs for cell balancing. ML techniques are broken down into reinforcement learning, unsupervised learning, and supervised learning.

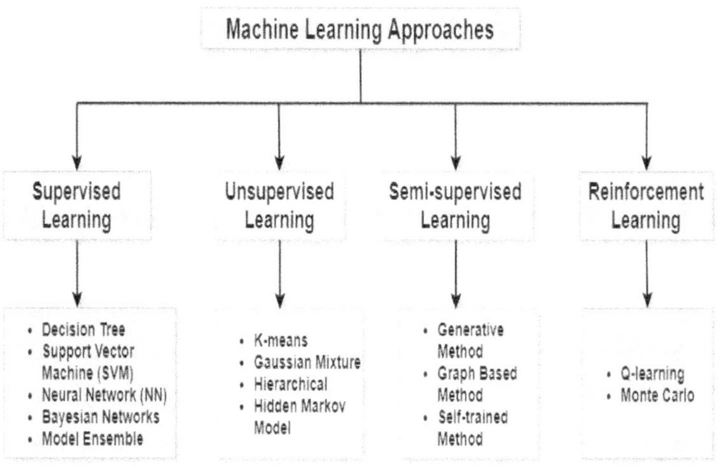

Figure 5.7 ML methodologies for BMS

The MATLAB (2019b) NN toolbox is used in the proposed approach to train, validate, and test the obtained battery data sets. The ML algorithm's flowchart is shown in Fig. 5.8 [Chandran, 2021]. Input parameters, feature extraction, and ML techniques are used to estimate the value of the balancing resistor. The ML algorithm is trained using the known data to create a charge curve forecasting model. Based on the three crucial factors of battery voltage, charging current, and battery temperature, the ML algorithm decides which balancing resistor to use.

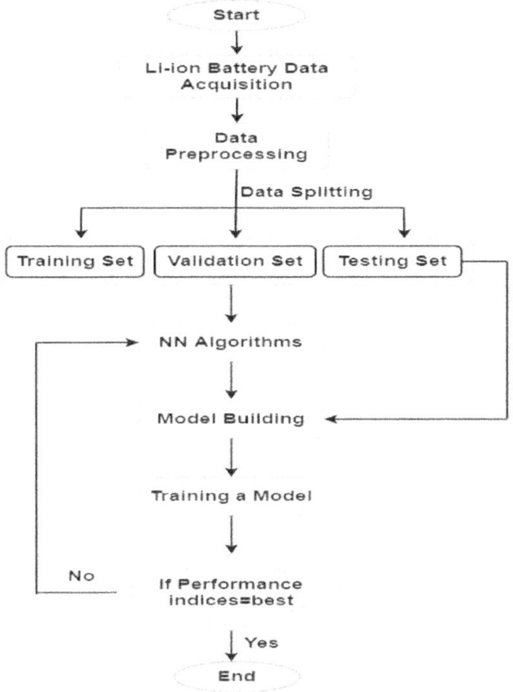

Figure 5.8 Flowchart of ML algorithm

5.3.1 Back propagation neural networks (BPNN)

A parallel processing method called BPNNs uses input/output data set patterns to learn the behaviour of the system. The artificial neurons' sigmoid activation function gives the network non-linear characteristics. The BPNN nodes are often trained using stochastic gradient descent and back-propagation techniques. Using regression analysis and mean square error (MSE) approaches, the BPNN fitting algorithm builds and trains a network for the input data. Because of its quick training and high accuracy, the Levenberg-Marquardt backpropagation algorithm is used to train the BPNN. Levenberg-Marquardt is an algorithm that employs backpropagation, which is a method for optimization. The bias and weight factors can be adjusted with the use of the Levenberg-Marquardt method. The Jacobian matrix of performance functions is then calculated using the backpropagation process while taking the bias and weight variables into account. The measured voltage, charging current, and temperature are displayed in inputs 1, 2, and 3, which are gathered from the corresponding battery sensors. Cell voltage, charging current, and pack temperature from the experimental balancing data (10,672) are provided in as inputs. To choose the value and duration of the balancing resistor, the output of the ML-based model is used. The accuracy of the balancing system is determined by the algorithm, which employs 20 hidden neuron layers.

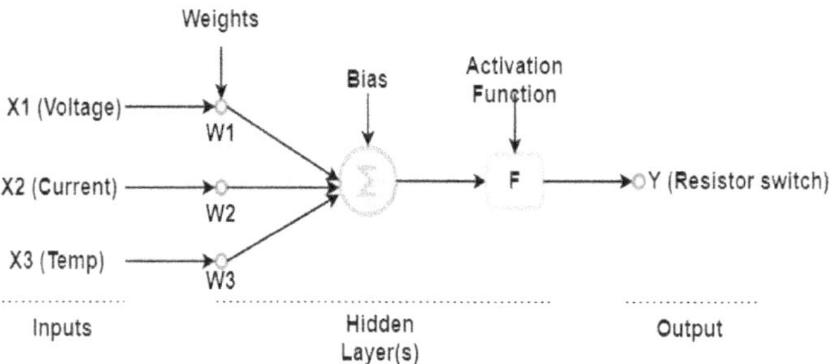

Figure 5.9 Layers of NN model

In this work, balancing resistors are chosen using a three-layer back propagation neural network model, as shown in Fig. 5.9. The input variables are characterized in the first layer, and the output variables are characterised in the third layer. The second layer determines the accuracy of the model outcome, which has one or more hidden neurons. The following steps explain the algorithm steps in detail [Hannan, 2018].

Step 1: Weight (w) and bias (θ) of the random variables are initialized

Step 2: The hidden layer sigmoid function defined in Eqn. (5.2)

$$F(net) = \frac{1}{1+e^{(-net)}} \tag{5.2}$$

For input q, the jth input layer holding $x_{q,j}$. Total input to the hidden layer (ith node) is given in Eqn. (5.3),

$$net_i = \sum_{j=0}^{n} w_{i,j}\, x_{q,j} + \theta_{i,j} \tag{5.3}$$

Where, $\theta_{i,j}$ and $w_{i,j}$ represent the bias and weight from the input to the hidden layer. The ith node hidden layer output as shown in Eqn. (5.4),

$$x_{q,i} = S_i\left(\sum_{j=0}^{n} w_{i,j}\, x_{q,j} + \theta_{i,j}\right) \tag{5.4}$$

S_i is the activation function. The total input to the output layer pth node is shown in Eqn. (5.5),

$$net_p = \sum_i w_{p,i} x_{q,p} + \theta_{p,i} \tag{5.5}$$

Where, $\theta_{p,i}$ and $w_{p,i}$, are the bias and weight to the output layer from the hidden layer. The output layer pth node output is shown in Eqn. (5.6),

$$SW_{q,p} = S_p\left(\sum_i w_{p,i} x_{q,p} + \theta_{p,i}\right) \tag{5.6}$$

Step 3: The error computed and backward propagated to the hidden layer from the output layer, and the output layer error computed from Eqn. (5.7)

$$\varepsilon_p = S_p(1 - S_p)(A_p - S_p) \tag{5.7}$$

A_p – is the actual output, the hidden layer error is calculated from Eqn. (5.8)

$$\varepsilon_i = S_i(1 - S_i)\varepsilon_p w_{p,i} \tag{5.8}$$

Step 4: Biases and errors are updated. The following Eqn. (5.9 - 5.12) update the weights

$$\Delta w_{p,i} = \beta \varepsilon_p S_i \tag{5.9}$$

$$w_{p,i} = w_{p,i} + \Delta w_{p,i} \tag{5.10}$$

$$\Delta w_{ij} = \beta \varepsilon_i x_q \tag{5.11}$$

$$w_{i,j} = w_{i,j} + \Delta w_{i,j} \tag{5.12}$$

Where β – is the learning rate. From the following Eqns. (5.13 - 5.16) biases updated

$$\Delta \theta_{p,i} = \beta \varepsilon_p \tag{5.13}$$

$$\theta_{p,i} = \theta_{p,i} + \Delta \theta_{p,i} \tag{5.14}$$

$$\Delta \theta_{ij} = \beta \varepsilon_i \tag{5.15}$$

$$\theta_{i,j} = \theta_{i,j} + \Delta \theta_{i,j} \tag{5.16}$$

For the test, an NMC cylindrical cell has been chosen. In Chapter 3, statistics on the most important battery parameters—current, voltage, and temperature—that have a big impact on balancing performance are gathered and described. Various cell imbalance situations, charging C-rates, and ambient temperatures (45°C, 20°C, and 0°C) are used during the test.

To decrease power loss and improve thermal efficiency, ML is used to speed up the balancing process and provide a customised design for balancing control. The battery charge current, cell temperature, and cell voltages are the first three inputs for the algorithm that was created. To choose the resistors, the controller output modifies the pulse width modulation (PWM) changing value of the switch. The ML method further maps the voltage, current, and temperature values into the relevant switch ON/OFF states for each cell. If the updated battery voltage is lower than the voltage values of other cells in the pack, balancing is halted. Under various levels of voltage imbalance, charging C-rates, and ambient temperatures, cell balancing behaviour during charging was observed, and the data were recorded. The resulting model is trained and tested using the 70% and 30% data sets, which have the best hidden layer neurons and learning rates. The maximum epochs taken into account in this investigation are 1000, and the performance objective is set at 0.000001. The outcomes of the BPNN passive balancing model are then contrasted with those of the traditional passive approach.

5.3.2 Radial basis neural network (RBNN)

As shown in Fig. 5.10, RBNN consists of a single hidden layer, an input layer, and an output layer. RBF networks typically consist of three steps: choosing the network size, choosing the initial parameters (widths and centres), and training the neural networks. Since Gaussian functions are typically employed, the radial distance is defined as r = ||x- t||, where t is a receptor. The definition of the sum squared distance between each cluster's closest samples and each receptor is given in Eqn (5.17),

$$\frac{1}{N} * \|x - t\|^2 \tag{5.17}$$

N = number of input values

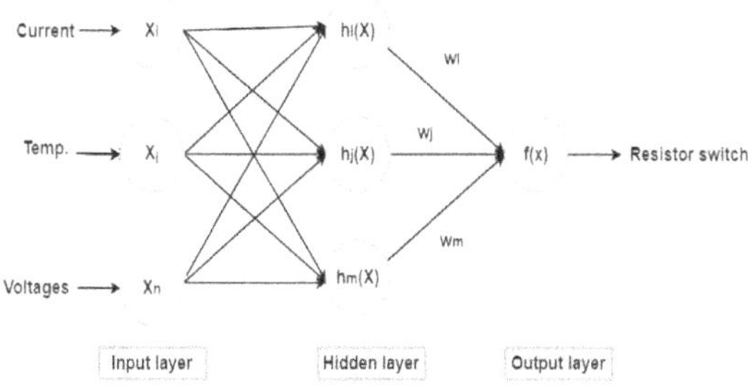

Figure 5.10 Layers of RBNN model

$$f(x) = \sum_{j=1}^{m} w_j \, h_j \, (x) \qquad\qquad (5.18)$$

$$h(x) = exp\left[-\frac{(x-c)^2}{r^2}\right] \qquad\qquad (5.19)$$

The output and hidden layer of the RBNN network are displayed in Equations (5.18) and (5.19). A collection of non-linear radial basis transformation operations are carried out by each hidden layer node. Using backpropagation techniques, stochastic approximation, and the clustering algorithm—, which also identifies cluster centers— hidden layers, are taught in the initial stage of training. For each hidden layer node, the receptors and the spread of the radial basis function must be identified. The weighting vectors between the hidden and output layers are updated in the second stage. In the buried layer, every node executes a transformation basis function. The output layer is then given the linear combination of hidden layer functions. The receptors are the output clusters. The final interpretation of the training step is the input vector projected onto the altered space. The RBNN design is depicted in Figure 5.11 with N inputs, K hidden layers, and M outputs [Xie, 2011]. The RBF algorithm includes the following steps: First layer (input layer) computation Eqn provides the input weights w^h that weigh the input vector x at the hidden unit k. (5.20).

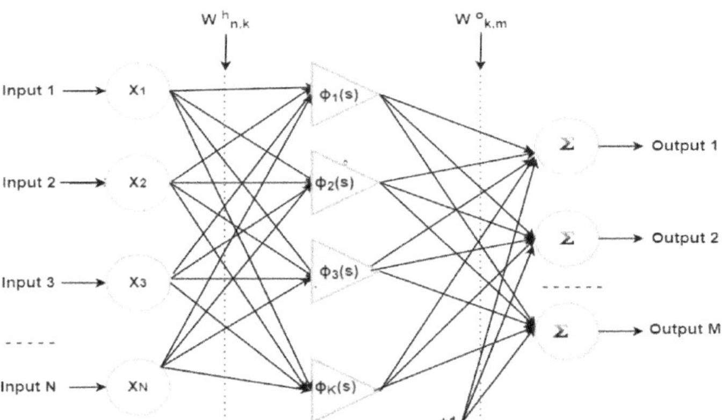

Figure 5.11 Generalized RBF network

$$S_l = \left[x_1 w_{1,k}^h, x_2 w_{2,k}^h \dots x_n w_{n,k}^h \dots x_N w_{N,k}^h \right] \qquad (5.20)$$

Where k - hidden units index, n - input index, $w_{n,k}^h$ is the input weight between n and k, x_n is the nth input.

Second layer computation: The hidden unit k output is derived from Eqn. (5.21),

$$\varphi_k(S_k) = exp\left[-\frac{\|S_k - C_k\|^2}{\sigma_k} \right] \qquad (5.21)$$

Gaussian function is the generally preferred activation function $\varphi_k(S_k)$, σ_k - hidden unit k^{th} width and C_k - hidden unit k^{th} centre.

Third layer computation: The output m of the network is derived from Eqn. (5.22),

$$O_m = \sum_{k=1}^{K} \varphi_k(S_k) w_{k,m}^o + w_{O,m}^o \qquad (5.22)$$

Where m - output index, $w_{O,m}^o$ - output unit m^{th} bias weight, $w_{k,m}^o$ - output weight between output unit m and hidden unit k. The model outputs are evaluated based on the mean square error given by Eqn. (5.23),

$$E = \frac{1}{q}\frac{1}{M} \sum_{q=1}^{Q} \sum_{m=1}^{M} e_{q,m}^2 \qquad (5.23)$$

Where e $_{(q,m)}$ is the error at output m when training pattern q, computed as the difference between the corresponding desired output and actual output, and q is the pattern index (1 to Q), m is the output index (1 to M). The hardware implementation of the passive cell balancing system produces data sets of inputs that include battery temperature, charging current, and cell voltages. The passive balancing controller was used to measure the voltage and current during charging for various levels of cell voltage imbalance at varied charging currents and temperatures. 10,671 training patterns and 1,199 testing patterns are included in the data.

5.3.3 Long short term memory (LSTM)

For issues with time series prediction, LSTM is a modernised version of RNN [Gers, 2000]. The backpropagation training algorithm solves the vanishing gradient issues over time. Instead of using neurons, which are utilised to solve difficult sequential issues, it uses memory blocks connected by layers. As illustrated in Fig. 5.12, the LSTM network model has an input gate, forgetting gate, and an output gate that considerably increase the LSTM's memory capacity in multi-step time series prediction problems. The LSTM approach was used to learn the data obtained from the real-time cell balancing implementation. Comparisons are made between the results of learned and actual cell balance. Python Tensor flow and Keras package is used for programme learning and writing.

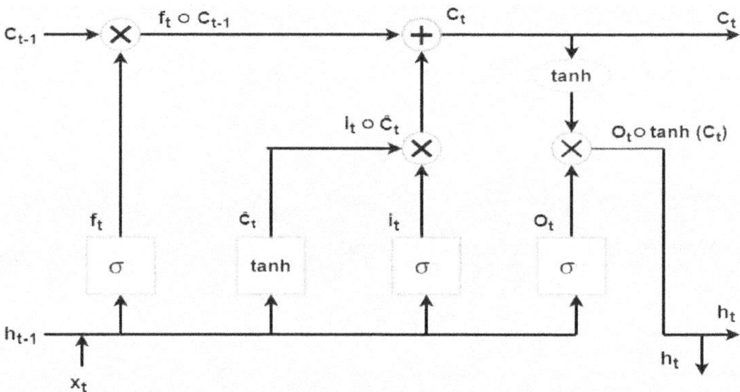

Figure 5.12 LSTM network architecture

The long term memory is called a cell state. Cell state equation is shown in Eqn. (5.24).

$$C_t = i_t \circ \hat{c}_t + f_t \circ C_{t-1} \tag{5.24}$$

From Eqn. (5.24), by multiplying with the forget gate, the information of the previous cell state (C_{t-1}) is removed, and new information is added to the input gates output.

The remember vector is called a forget gate, as shown in equation (5.25).

$$f_t = \sigma\big(w_f[h_{t-1}, x_t] + b_t\big) \tag{5.25}$$

By multiplying a matrix position by 0 to determine which information needs to be erased, the forget gate determines the state of the cell. If the forget gate output is 1, the cell remains in that state. The sigmoid function is applied to the weighted input and previous hidden state using Eqn. (5.26). Which data should be stored in long term memory is determined by the save vector (input gate).

$$i_t = \sigma(b_i + w_i[h_{t-1}, x_t]) \tag{5.26}$$

Input modulation gate equation is represented in Eqn. (5.27),

$$\hat{c}_t = tanh(w_c[h_{t-1}, x_t] + b_c) \tag{5.27}$$

It makes use of the [-1, 1]-ranged tanh activation function to let the cell state erase the memory. The output vector is often referred to as the focus vector. It chooses which data should be transferred to the following hidden layer as specified by Eqn (5.28),

$$O_t = \sigma(b_o + w_o[h_{t-1}, x_t]) \tag{5.28}$$

The working memory is the hidden state. It chooses which pieces of information should be added to the subsequent order specified by Eqn (5.29).

$$h_t = O_t \,^\circ tanh\,(C_t) \tag{5.29}$$

Note: In the proposed approach, in order to run the machine learning algorithm, the micro controller and memory resources to be adequately planned.

5.4 Results and discussions

Matlab-Simscape 2019b captured the passive simulation. Voltage measurement and switching blocks make up the system's two basic building blocks. To learn the balancing features, the simulation was run using a different set of charge currents of 0.2C, 0.5C, and 1C. Three NMC cells, each with a 3.35Ah capacity and nominal, discharging, and charging cut-off voltages of 3.6, 2.7, and 4.2V, respectively, are used in the simulation. Better balancing performance is achieved with a smaller differential voltage (ΔV) between the cells in the pack. The voltage differential between brand-new and older Li-ion cells utilised in EVs is smaller. The voltage difference between cells 1, 2, and 3 is fixed between 25 mV and 275 mV in order to investigate the balancing time, current required, and maximum voltage deviation between cells under fast charging conditions. Since the power loss is more significant for higher balancing currents, designs, and resistor selection for higher power loss to be taken into consideration, Table 5.1 shows the balancing current, time, and power loss of the cell at different C-rates with respect to variable resistors to balance the cells within the required charging duration.

Table 5.1 Power loss vs. balancing resistance

Cell	C-rate	R_B (Ω)	I_B (mA)	BT (mins)	Power Loss in R_B (W)
	0.2C (0.7A)	36	100	175	0.35
	0.5 C (1.75A)	12	300	65	1.04
C2 (min. ΔV)	1.0 C (3.5A)	9	400	56	1.40
	0.2C (0.7A)	7	500	250	2.0
	0.5 C (1.75A)	3	1286	117	5.2
C3 (max. ΔV)	1.0 C (3.5A)	1	4000	54	12.5

The balancing time based on the difference in voltage between the cells is shown in Table 5.2. Lower V requires the shortest time to balance. The balanced performance, however, fell short of expectations. Based on the user's needs, the cells' balancing threshold voltage (ΔV) adjustment.

Table 5.2 Balancing time at different ΔV

Balancing Threshold Voltage (mV)	Balancing Time (mins)
10	58
20	38
30	27
40	21

When compared to the established balancing techniques [Amin, 2017], [Valchev, 2018], the suggested balancing concept can effectively reduce balancing time. Shunt resistors placed in a scattered pattern on the printed circuit board (PCB) keep the cell temperature within the safe range since the power loss across the balancing resistor is greatest when the balancing current is high, which causes the pack's temperature to rise.

In this distinctive statistical ML technique, the outcomes are benchmarked and compared using various ML models. The voltage differential between the first and third cells is kept larger than that of the second cell. Table 5 displays the simulation scheme. 3. With regard to cells 2 and 3, cell 1's initial voltage differences are 55 mV and 101 mV, respectively. There is a 36 mV voltage differential between cells 2 and 3. Passive balancing primarily takes place in the start or last stages of charging due to the non-linear relationship between battery voltage and SOC [Prabhakar, 2016], [Shabarish, 2020]. Balancing is completed toward the end of charging using the final voltage-based technique. Furthermore, regardless of the charging C-rate, the balancing time and power loss through the balancing resistors are the same.

Table 5. 3 Variable voltage, C-rate and balancing threshold scheme

Cells	Initial Voltage (V)	C-rate	Balancing threshold (mV)
Cell1	3.528V	0.2C, 0.5C, 1C	30, 50, 100
Cell2	3.427	0.2C, 0.5C, 1C	30, 50, 100
Cell3	3.473	0.2C, 0.5C, 1C	30, 50, 100

Using both traditional and suggested ML-based cell balancing strategies, the balancing current, cell voltage, balancing time, temperature, and other parameters were measured at each charging C-rate and balancing threshold. All of the crucial factors have been taken into account when creating the test design for the passive balancing system. The results of the performance analysis of the cell balancing system are plotted. capacity ratio balance [Wei, 2017],

$$\alpha = \frac{C_{bal}}{C} \times 100\% \tag{5.30}$$

C_{bal} reflects the capacity used (Ah) during balancing, and C represents the rated capacity (Ah).

Balancing current ratio,

$$\beta = \frac{I_{bal}}{I} \times 100\% \tag{5.31}$$

I_{bal} stands for the balancing current (A), while I denotes the battery's charging current (A). Balancing time ratio,

$$\mu = \frac{t_{bal}}{t} \times 100\% \tag{5.32}$$

t_{bal} denotes the balancing time (s), and t is the battery's charging period. According to Fig. 5.13, the larger the balancing capacity ratio, the bigger the voltage difference between cells. The strong non-linear behaviour of the Li-ion battery at the early and end stages of charging is the cause of the non-linearity between the capacity ratio and 1V.

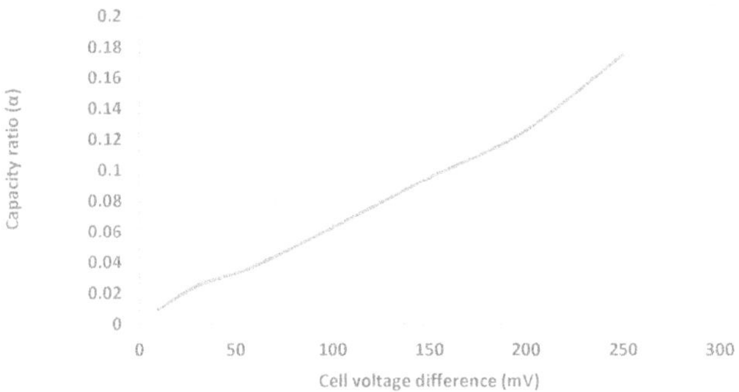

Figure 5.13 Cell capacity ratio vs. voltage difference

As depicted in Fig. 5.14, the balancing time ratio rises as the battery's current ratio falls. As the voltage variation between the cells increases, it is seen that the balancing time ratio rises.

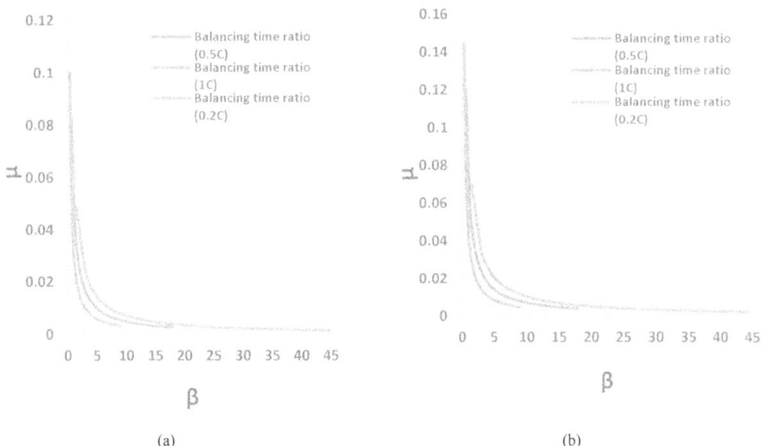

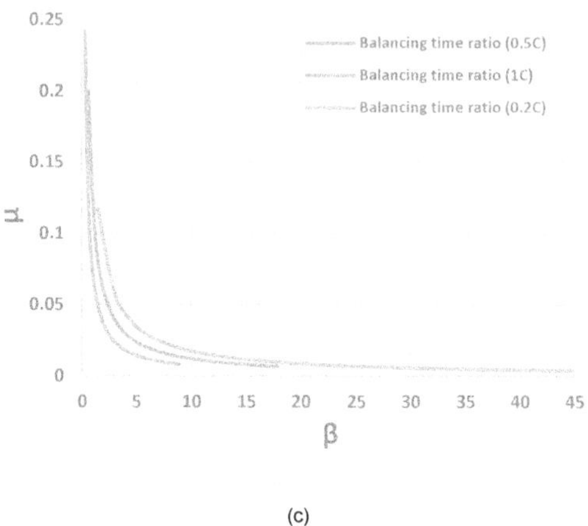

(c)

Figure 5.14 β vs. μ, (a) at 30mV initial ΔV (b) 50 mV initial ΔV (c) 100 mV initial ΔV

A suitable increase in the balancing current ratio can help to improve the balancing impact in real-time applications since the balancing current ratio and the balancing time ratio are inversely proportional. In order to anticipate the resistor value under various cell balancing scenarios, the learning-prediction based NN model selects the balancing resistor from the set of balancing data. Various charge C-rates are used in the experiments for online balancing and charging. The feed-forward neural network trained with the Levenberg-Marquardt backpropagation technique is fed the balancing data. With a hidden layer size of 20, the system's data set consists of 7,470 training, 1,601 testing, and 1,601 validation patterns. Fig. 5.15 shows the training and testing MSE of 0.0046 and 0.0048 along with the best validation performance MSE of 0.0045. In the suggested balancing system, as illustrated in Fig.5.16, cell balancing is carried out using the developed Simulink NN model (which makes use of the Machine Learning method).

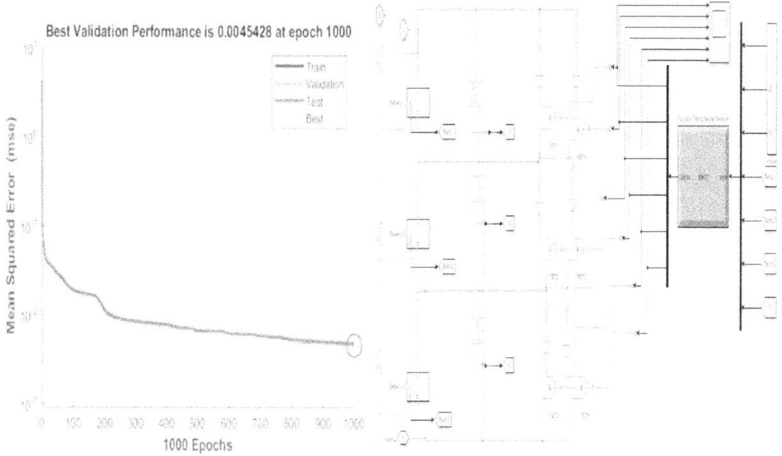

Figure 5.15 Proposed NN model performance **Figure 5.16 Proposed model in Matlab (2019b)**

The results of the simulation demonstrate that the passive system has reached its ideal balance parameter values. In the typical method, cell1 and cell3 require 165 and 58 minutes, respectively, to balance, regardless of the charging C-rate. The suggested ML-based solution takes a lot less time and has less power loss than traditional cell balancing. From the simulation resulting waveforms of Fig. 5.17 to 5.20, the balancing time, power loss, and temperature fluctuations of conventional and proposed methods are investigated. Table 5.3 shows that the voltages of cells 1 and 3 are higher than those of cells 2; as a result, cells 1 and 3 are in balancing mode while cells 2 are in normal mode. As shown in Fig. 5.17(a), the drive signal for cells 1 and 3 is thus turned on. Figure 5.17 (b) displays the balancing current, battery temperature, and power loss across the balancing resistors (d). Temperatures for Cells 1, 2, and 3 are shown as T1, T2, and T3, respectively. Cell 1 is in the balancing phase for approximately 9,868 seconds. In order to demonstrate the voltage drop generated by the balancing resistance when the current flows through the balancing circuit, this phenomena is represented in Fig. 5.17(b) and (d). Finally, all three cells are in the normal mode after 13,994 seconds since the voltage deviations are much lower. The voltage variances are significant at the start of charging and balanced towards the end, as seen in Fig. 5.17(e). In this setup, charging was used for the test rather than idle state. As a result, the data indicate an increase in voltage rather than a decrease. A 100 mA or so balancing current is used. The battery temperature is under 40 °C, and the power loss is under 0.4 watts. It takes 165 minutes to balance cell 1 and get the voltage difference from 101 mV to less than 30 mV. To balance the voltage difference of 46 mV with cell 2 and 55 mV with cell 1, it takes cell3 85 minutes.

According to Figs. 5.17 and 5.19, the time required by typical cell balancing systems to balance the cells at 0.5C and 1C charging is nearly constant. The only issue is the short amount of time to balance because balancing is done towards the end of charging, when the charging current is at its lowest. Figs. 5.18 and 5.20 depict the proposed balancing plan. The varied degrees of cell imbalance and charging conditions are taken into account when balancing. When the ΔV is between 30 and 50 mV, resistor-1 (33Ω), which is placed across the unbalanced cell, is switched ON. When the voltage is between 50 and 100 mV, resistor-2 (25Ω), as well as both resistors, are turned ON.

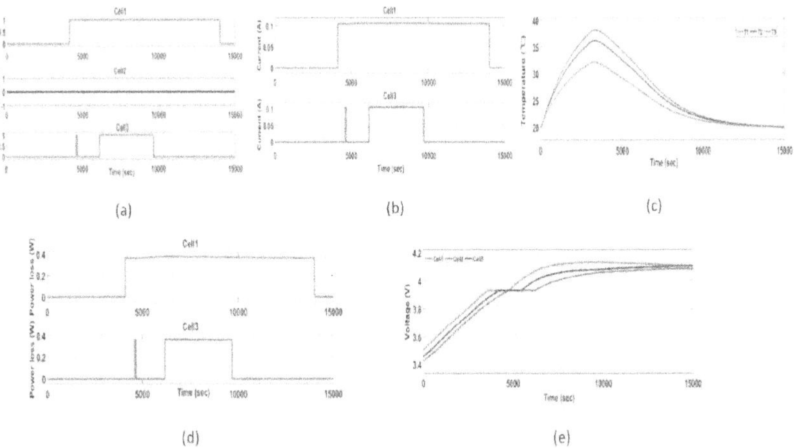

Figure 5.17 Conventional system charging at 0.5C a) switching signal (b) current through the balancing resistors (c) battery temperature (d) balancing resistor power loss (e) cell voltages

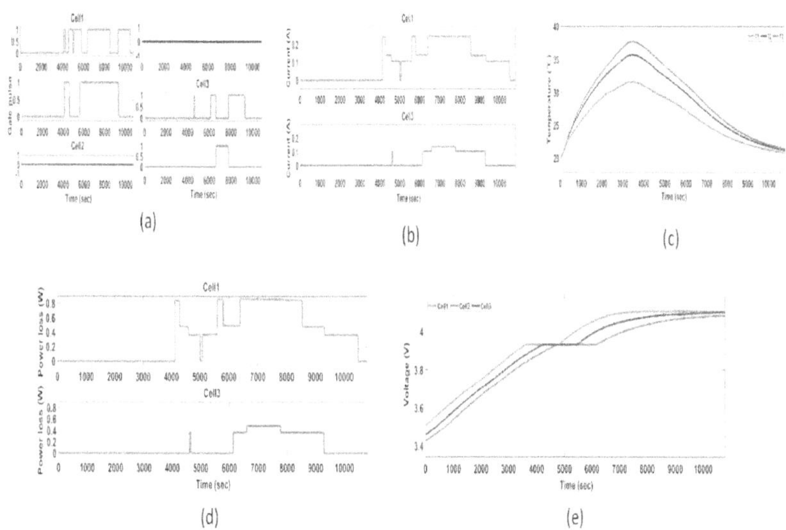

Figure 5.18 Proposed system charging at 0.5C a) switching signal (b) current through the balancing resistors (c) battery temperature (d) balancing resistor power loss (e) cell voltages

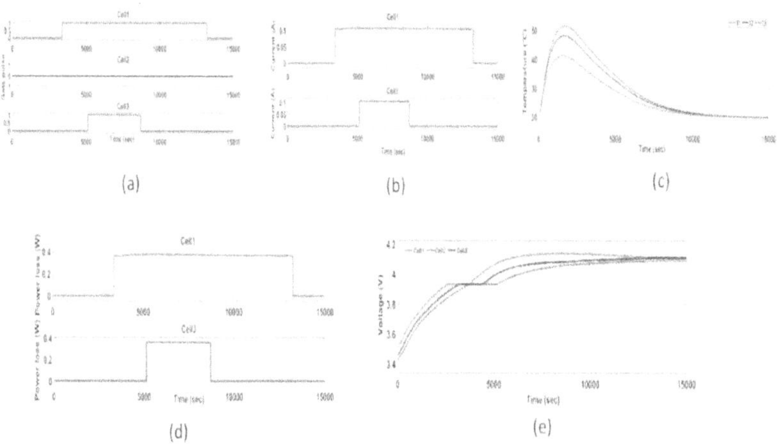

Figure 5.19 Conventional system charging at 1C a) switching signal (b) current through the balancing resistors (c) battery temperature (d) balancing resistor power loss (e) cell voltages

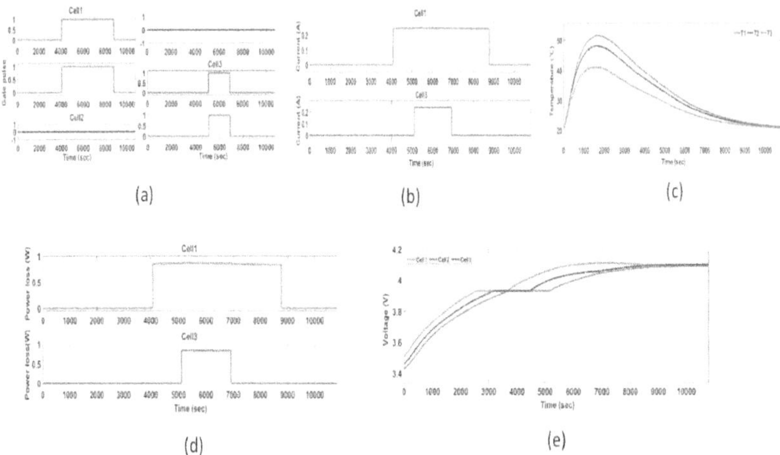

(a) (b) (c)

(d) (e)

Figure 5.20 Proposed system charging at 1C a) switching signal (b) current through the balancing resistors (c) battery temperature (d) balancing resistor power loss (e) cell voltages

As a result, the balancing process is accelerated, which enhances the balancing impact. Figure 5.18 shows that when V is over 100 mV, both resistors are turned on until V drops below 100 mV, at which point resistor-2 is turned on (a). In the suggested strategy, cells 1 and 3 require 101 and 77 minutes, respectively, to reach the balancing threshold. At the same time, the overall power loss and battery temperature do not significantly increase. As long as the battery temperature is within safe ranges, both resistors are turned ON for a higher C-rate, as shown in Fig. 5.20. Cells 1 and 3 stay in the balancing mode at 4,680s and 1,800s respectively before switching to regular mode. In Table 5.4, the proposed system is compared to the conventional system in terms of balancing time, energy loss, and battery temperature. The proposed approach exhibits a significant shift in the balancing time, notably at higher C-rates. Keeping a healthy temperature range during cell balancing will lengthen the battery's lifespan.

124

Table 5.4 Proposed passive system comparison

	CPBS (0.5C)		PPBS (0.5C)		CPBS (1C)		PPBS (1C)	
Cell. No	C1	C3	C1	C3	C1	C3	C1	C3
BT (min)	165	85	101	77	165	60	78	30
EL (Wh)	1.1	0.4	1.0	0.36	1.1	0.36	1.1	0.4
T_B (°C)		38		37.7		51.5		51.7

* PPBS: proposed passive balancing system, CPBS: conventional passive balancing system, EL: energy loss, BT: balancing time, T_B: battery temperature

To compare performance, the same data set is trained using RBNN and LSTM. The best validation performance is attained at two epochs in RBNN with an MSE of 0.0717 due to their quicker learning rate, as illustrated in Fig. 5.21. The convergence of optimization targets happens significantly faster because there is only one hidden layer.

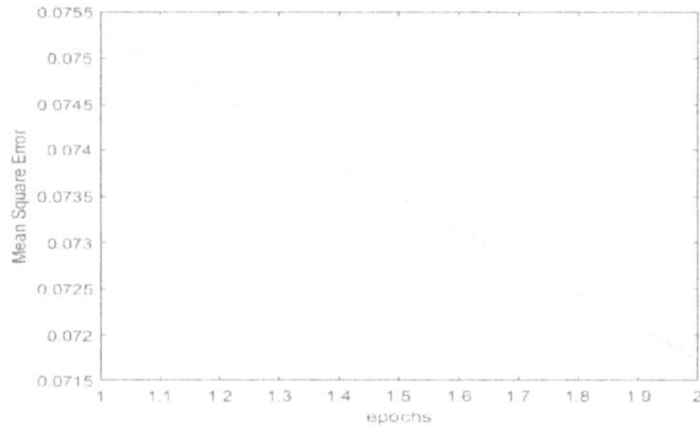

Figure 5.21 Best validation performance indices of RBNN at two epoch

The time series data are transformed into sequential data with time steps using LSTM. The time steps show how many past steps must be taken into account in order to forecast the output at a later period. Six data properties are predicted using various execution-generated performance indices, output graphs, and timestamps. Data on cell balancing is gathered using a temperature range of 0°C to 45°C with charging currents of 0.2C, 0.5C, and 1C. Temperature, cell voltages, charging current, and charging time are among the statistics. 10,642 pieces of information are used to train the algorithm, of which 70% are used for training and 30% for testing. Three dimensions make up LSTM values: samples, time steps, and features. LSTM training is a crucial stage in this process. Twenty neurons make up the first layer of the LSTM, and the input shape consists of seven features taken into account across 30 time steps. We employ the Adam variant of stochastic gradient descent together with the MAE and MSE loss functions. With a batch size of 20, the model can fit 20 training epochs. The number of epochs required to predict the data with time and losses is shown in Fig. 5.22. It's interesting to note that after 13 epochs, the model's fit to the training data causes the testing loss to marginally fall below the training loss. The model then exhibits a perfect fit, beyond which the model training could be terminated at the inflection point and the quantity of training data increased.

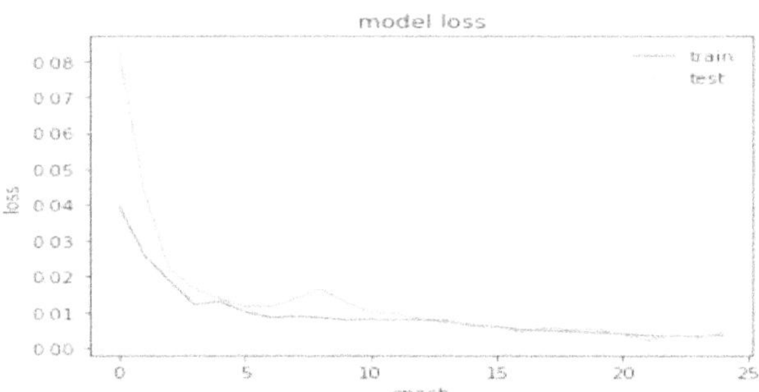

Figure 5.22 Multivariate LSTM testing and training loss

While RMSE and RMAE are computed after on an inverse scaled target and reflect actual performance, loss is computed during training on a scaled target. Fig. 5.23 displays the switching state that was obtained at each algorithmic phase. For 500 time steps, the switching state is anticipated. The projected data matches the real data with a 0.01 acceptable error range, demonstrating the accuracy of the predicted values.

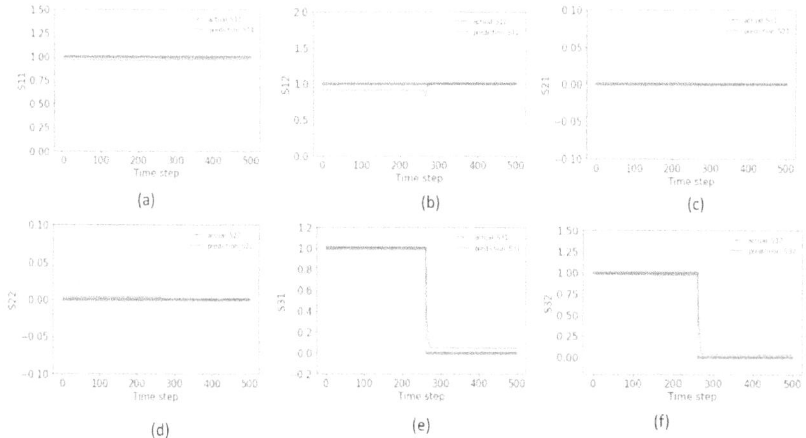

Figure 5.23 Actual and predicted values of balancing switching states using thirty previous time steps at twenty epochs

The performance error is displayed at various time stages in Table 5.5. As the number of time steps rises, it is seen that the MSE and MAE values rise as well. Not enough time is given for the increased network capacity to accommodate the data.

Table 5.5 LSTM model error metrics at different time steps

Error metrics	30 Time steps	50 Time steps	100 Time steps	150 Time steps
MSE	0.010	0.013	0.015	0.016
MAE	0.054	0.076	0.090	0.097

Increasing repetition and epochs would help to decrease performance errors. At the conclusion of each epoch, the testing and training losses are recorded. Finally, the test data set's MSE and MAE are shown. As seen in Fig. 5.24, the model obtains a low MSE and MAE as the number of epochs rises. The testing is conducted fifty times for each model, and Figure 5.25 displays the net trajectory of testing/training errors.

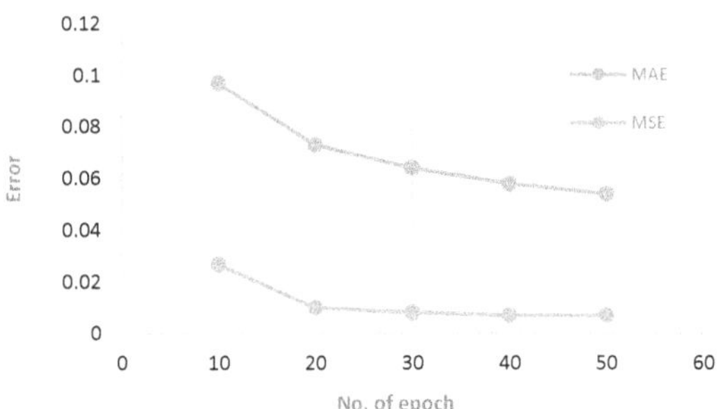

Figure 5.24 Errors when epoch increases

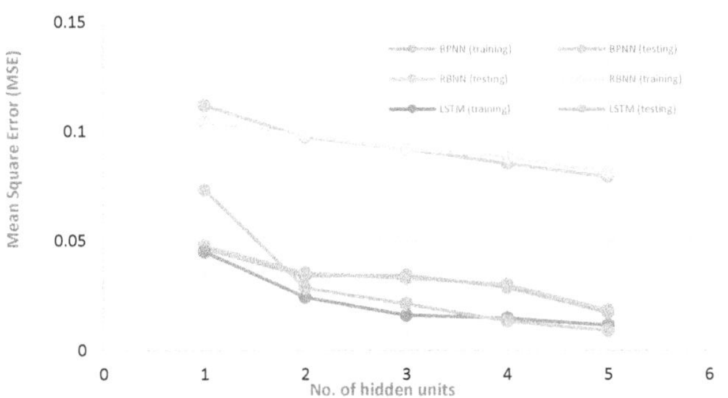

Figure 5.25 Comparison of BPNN, RBNN, and LSTM model: training, testing error vs. the number of hidden units

128

The LSTM obtained lesser training/testing error than BPNN and RBNN. As the number of hidden layers increases, all three models perform better due to decreased overall error. The overall MSE, RMSE, and MAE are calculated from the repeated testing for the three models, shown in Table 5.6.

Table 5.6 MAE, MSE and RMSE of BPNN, RBNN and LSTM model

Model	MAE	MSE	RMSE
BPNN	0.1276	0.0310	0.1760
RBNN	0.1930	0.0717	0.2677
LSTM	0.0660	0.0090	0.095

The performance of BPNN is superior to RBNN for less hidden units. The BPNN takes longer to converge than the RBNN, but it performs well. The proposed strategy is better suited if cell unreliability is at its highest, particularly for ageing cells, and charging time is at its lowest (rapid charging), where the user has a shorter battery charge time. The non-linear battery characteristics allow the ML-based cell balancing approach to properly anticipate the degree of cell imbalance and choose the appropriate balancing resistor. Only three cell and three resistor combinations are used in the simulation. But more resistors can be chosen dependent on charging and balancing parameters like charging time, battery SOA, available space on the BMS board, etc.

5.5 Conclusion

The main goal of this work is to emulate the passive balancing properties of serially connected Li-ion cells while they are being charged quickly with direct current (DC). The simulation of passive balancing using shunt resistor circuits has been completed. The balancing process moves along much more quickly with a higher balancing current. Therefore, it is appropriate for quick DC charging. Although it takes longer to balance, a pack with less voltage variation between the cells provides better balancing performance. In order to balance, a higher balancing current would require more shunt resistors, which would require more PCB space for heat dissipation. Therefore, the shunt resistor must be positioned correctly to withstand the thermal load. The most effective method of improving the passive balancing mechanism is to provide a longer battery pack lifetime and shorter balancing time due to the high cost and design complexity of the active balancing system.

Comprehensive ML algorithms are used to balance the cells of the Li-ion battery pack precisely by estimating optimal resistor values, which take into account factors like balancing time, power loss, temperature rise, and operating constraints and further optimise the balancing currents in the system. In order to reduce the computational complexity of a big battery pack, it is more practical to implement the variable resistor-based passive cell balancing technique in real-time. Due to their superior handling of non-linear data, BPNN, RBNN, and LSTM are used to test the performance of the suggested technique. Numerous example outcomes show how efficient the suggested ideal cell equalising control approach is. The LSTM performs better with an MAE of 0.0660 because its kernel can extract complex time-structured, sequential data. With 0.1276 MAE, BPNN technique outperforms RBNN in performance. The proposed system's results show a faster balancing speed and a reduced temperature rise when compared to a conventional technique. To sum up, the probability distribution in the suggested ML-based method further promotes improvement in the cell balance estimation.

CHAPTER 6
Conclusions and Future Scope

6.1 Introduction

This section represents the conclusion of the work carried out for this research and its future scope.

6.2 Conclusion

This work's key contribution is to enhance the effectiveness of the passive cell balancing system using circuit experiments and simulations based on the 3RC equivalent circuit model for Li-ion (NMC) cells that take into account the voltage variances of the cell's lowest and maximum. This research's main goal is to model the passive balancing properties of serially connected Li-ion cells under a fast charging scenario while optimising and fine-tuning the passive balancing algorithm for efficient thermal management of the parallel resistor array.

Five different battery modelling techniques that are effective and trustworthy for use in a variety of operating scenarios are explored in this paper. These models outline cutting-edge methods for evaluating and modelling batteries. The performance traits of the Li-ion battery have been assessed using RC ECM calculation. The findings demonstrate that the 3-RC model, which better captures the nonlinearity of the system than lower order modelling, establishes the optimal balancing. The model created in this study is scalable to BMS applications at the pack level and operates at the cell level. Therefore, for the given input current stimulus, 3-RC ECM provides a better level of cell characterisation. These battery model results can be used more effectively by BMS applications. As a result, the type of application and the designer's preference determine the best battery model. This study aids in choosing the best model for a specific situation. This study aids in choosing the best model for a specific situation.

There are several active balancing techniques presented. Based on the assessment, an inductor-based topology is chosen that can simultaneously change top, middle, and bottom cells with fewer parts than conventional inductor-based systems. Matlab-Simscape is used to model the sixteen-cell inductor-based active balancing system. Through simulation and hardware, the passive balancing approach of Li-ion batteries for EV applications is implemented and studied. The cells are balanced using a final voltage-based balancing method when the SOC is high. Results from the experiment are contrasted to those from the theoretical investigation. By taking into account the impact on power loss, system cost, and battery performance, the most important characteristics of the passive balancing system may be developed from the results of this experiment.

By adopting active cell balancing, the energy efficiency of the battery pack is raised by 1.86% for new cells and 9.17% for aged cells. Active balancing can increase the charge and discharge cycles by 0.1 to 0.5%. When compared to new cells (ΔV <30mV), the passive balancer requires greater time (4 hrs.) and significant energy dissipation (16 Wh). Additionally, a higher percentage of unbalanced cells in the pack is detected, which results in a large power loss (4.55 W). According to the analysis, with cells having a minimum ΔV (<30mV), the cost for the energy loss due to balancing is modest. Low power systems with closely matched cells, as those in E-bikes and E-rickshaws, can benefit from passive balancing. More series connected cells make up the battery pack for high-power applications (such as E-cars, E-trucks, etc.). Due to the potential for substantial imbalance, a high current active balancing topology is required. Fast charging is possible with a greater balancing current (>1A), which considerably lowers the balancing time. However, active balancing has a unique favourable, intricate, and expensive design. As a result, the battery pack's lifetime is increased and balancing time is decreased by this optimal concept to improve the passive balancing mechanism.

A high balancing current during balancing would call for more balancing resistors, which would demand a larger PCB surface to dissipate heat. To effectively manage the thermal load, balancing resistor placement is necessary. Despite taking more time to balance, the battery pack's decreased voltage difference between the cells provides good balancing performance. We evaluate the energy loss, battery temperature, and balancing time of the proposed ML-based system to the conventional balancing system. At a greater balancing current rate, the proposed methodology considerably shortens the balancing time. To extend the life of the battery, balancing must be done in a safe temperature range. Due to their superior handling of non-linear data, three important ML algorithms—LSTM, RBNN, and BPNN—are used to evaluate the performance of the suggested approach. The LSTM outperforms the other two models with MAE and MSE of 0.0660 and 0.0090, respectively. Despite the benefits, active balancing is expensive and requires a complicated circuit design. Therefore, this ideal suggestion for improving the passive balancing mechanism offers a longer battery pack lifetime and a shorter balancing time.

6.3 Future research orientations

Further research work in cell balancing may consider the higher-order voltage deviation in the balancing algorithm to fine-tune the system performance. Future research is trending towards including wide bandgap devices such as GaN and SiC to improve performance, circuit optimization, and intelligent techniques such as heuristic optimization, ANN and various machine learning approaches to improve the existing solutions. In order to run the AI/ML based techniques running in the BMS embedded system needs additional CPU power, memory and storage resources along with communication to the cloud, which can be taken for the future, scope of work. In addition, using cloud technology, the solution can be extended further to accommodate the vehicular data learned across multiple systems to provide an effective balancing, preventing abnormal failures. Optimal tuning of switching cycles while balancing multiple cells in a pack simultaneously against the loss can be considered for future work